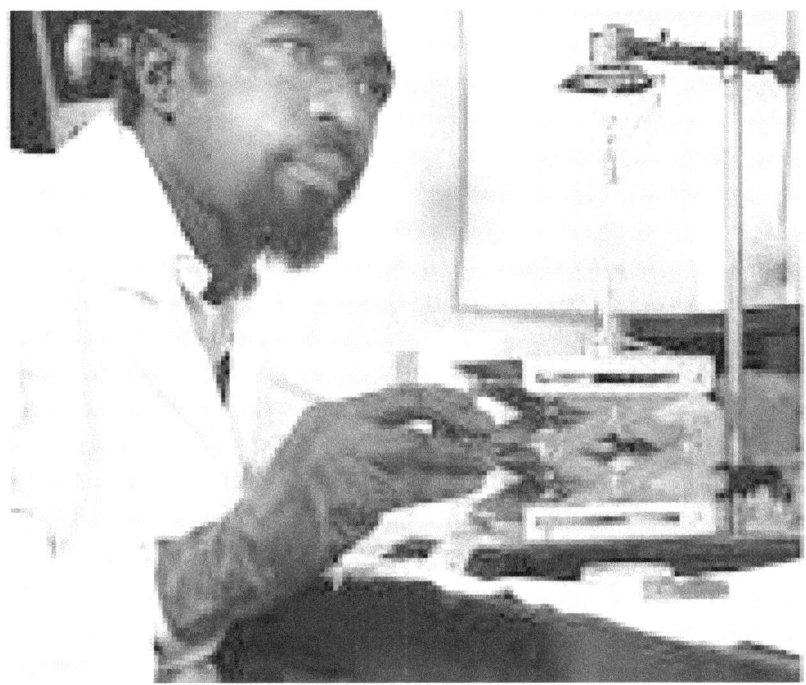

James Harris, a Lawrence Berkeley Laboratory (LBL) scientist and the first African American to play a key role in the search for new elements. He is credited with helping to discover rutherfordium and dubnium.

Global Contributions of Black People in Science and Technology

Past, Present, Future

Compilation by **Lee O.** Cherry
President and CEO
African Scientific Institute, USA
www.asi-org.net

Second Edition

Global Contributions to Advancements in Science and Technology by Black People in Africa and the African Diaspora

Past, Present, Future

© by the African Scientific Institute,
P.O. Box 20810, Oakland, CA 94620
www.asi-org.net
510-653-7027

No part of this book may be reproduced, stored in a retrieval system, or transmitted by any means, electronic, mechanical, photocopying, recording, or otherwise, without written permission from the African Scientific Institute, or its representative thereof.

ISBN 978-1-7359758-0-1

Acknowledgement

There has been so much documentation abouth Black history. It is very difficult to not repeat much of the same information that is covered in countless books and innumerable website sources.

My attempt in this book is to gather information and contain it in a short document about the contributions Black have made in the development of civilization with science and technology.

I worked to give credit to sources that I found. Please excuse any error I may have made in not quoting exact sources.

Foreword

Before You Start Reading This Book………..

1. Per Lewis H. Carlson:
"The evolution of the term "sciences" in Western civilization has burdened the concept with many connotations. Surely "science" implies as much in the realm of cultural arrogance as in the more proper sphere of empirical thought. The term as it now stands, cannot be appled fairly to the world of traditional Africa, or, for that matter, anywhere else beyond the restrictive bonds of Western civilization.

The term "science" remains to us, part of our definition and categorization of the world and the universe. The definitions of these elements in Africa (history) are totally alien to our own; it follows that our concepts of science cannot apply. A great deal has been done, for example, in the field of herbalogy by African; yet mostis still oral tradition and yet to become part of the Western pharmacopoeia."

2. Uniderstand that a human, even under slavery, is more than a physical being. He/she has spiritual, mental, and cultural attributes. When a person is transported from place to place, he/she brings knowledge with them.
3. The printing press wasn't invented in Germany until around 1440 by Johannes Gutenberg.
4. Before 1800, people were primarily living in agrarian based economies. The industrial revelution had not yet taken hold.

The Industrial Revolution marked a period of development in the latter half of the 18th century that transformed largely rural, agrarian societies in Europe and America into industrialized, urban ones. Goods that had once been painstakingly crafted by hand started to be produced in mass quantities by machines in factories, thanks to the introduction of new machines and techniques in textiles, iron making and other industries. Fueled by the game-changing use of steam power, the Industrial Revolution began in Britain and spread to the rest of the world, including the United States, by the 1830s and '40s. Modern historians often refer to this period as the First Industrial Revolution, to set it apart from a second period of industrialization that took place from the late 19th to early 20th centuries and saw rapid advances in the steel, electric and automobile industries.

5. Africa's Empires had started to collapse through conflicts and slave trades. But transported slaves had knowledge about agricultural development systems. They also had extensive knowledge about plants and their usages.

6. Public education for the masses of people did not take hold on a large scale until the 20th centrury. Knowledge was exchanged among the priviledged few. Rulers controlled who got educated and what they were educated about.

7. The enslaved Africans in the African Diaspora were not allowed to take credit for their advances in knowledge and inventions. Slave masters took credit for these attributes.

****

"Let no man define who you are and your destiny".

Throughout the history of mankind people have sought to better their lives through inventions. Many of these inventions came through trial and error. Many evolved from theoretical inquiry and mathematical modeling. Black people throughout the African Diaspora have contributed, are contributing, and shall continue to contribute to inventions and advancements in science and technology.

One of the most profound findings in researching information about people of color and their inventions was the classification system devised to accurately document information about Black people. This system was used as a decoy to keep black people from identifying one another as members of the same race. It was common practice to classify Black people under a host of different categories: Aborigine, African, Afro, Black, Caribbean, Colored, Dravidian, Khaffir, Moors, Mulatto, Native African, Negro, Octoroon, Quadroon, Slave, South African, West Indian,.....etc.[34]

Through exhaustive work, researchers have uncovered the contributions to advancements in science and technology made by

Black people throughout the African Diaspora. As I write this paper, more such information is unfolding.

****

Table of Contents

	Page
ANCIENT PAST	
Black Contributions from Africa ……………………...	2
THE VANISHING EVIDENCE OF CLASSICAL AFRICAN CIVILIZATIONS, by Prof. Manu Ampim	3
10 African Inventions That Changed the World …….	11
Northern Africa ………………………………………...	22
Western Africa …………………………………………	31
Central Africa …………………………………………	35
Eastern Africa …………………………………………	37
Southern Africa ………………………………………..	45
Black Contributions to Asia	
5 Ancient Black Civilizations That Were Not in Africa ………………………………………………..	47
Black Contributions to the Americas ………………..	67
PAST (0 AD – 1800)	94
African Empires ………………………………………	97
Northern Africa ………………………………………...	117
Western Africa …………………………………………	121
Central Africa …………………………………………	127
Eastern Africa …………………………………………	130
Southern Africa ……………………………………..……	138

	Page
PAST (1700 AD – 1875 AD) **PERIOD OF COLONIALIZERS' DISRUPTIONS**	148
Canada	173
United States	176
Latin America	178
Caribbean	186
Europe	188
Asia	189
CONTEMPORARY **(1900 AD – PRESENT)**	
Africa	202
Africa, its countries' natural resources and ASI Fellows	
African Diaspora	220
North America	
Canada	227
United States	229
Blacks in Latin America	
Mexico	236
Central America	238
South America	246
Caribbean	257
Europe	269
Asia	271
FUTURE	273
Black People of Africa and the African Diaspora Helping Each Other	279
CONCLUSION	294

BIBLIOGRAPHY	295
APPENDIX	310

Ancient Past

Black Contributions from Africa

Africa has a glorious history, from Egypt, Kush, Ghana, Mali, Ashanti, Songhai, Timbuktu Empires.

According to Lewis H. Carlson of Western Michigan University (*The Negro Impact On Weastern Civilization*, pgs 51 – 73);

Ancient Egyptians developed many of the basic skills of mathematics and geometry, that portion of North Africa contributed mightily both to the preservation and transmission of earlier learning. To count these elements among African scientific achievements, however, is to run afoul of two major problems. In the case of the Egyptians, the nature of the Pharaonic society and the gap of millennia between that time and the present have served to erase the names of the scientific "heroes" of the day. To seek the "inventor" of the pyramidpplied mathematics is akin to seeking the "inventor" of fire or the wheel. Western science, is founded on experimentation, using legions of people toiling for years seeking a single goal and placing individuals on pedestals.

The term "science" remains part of our definition and categorization of the world and universe. The definitions of these elements in Africa have been historically alien to Western civilization; it follows that our concepts of cannot apply. A great deal has been done, for example, in the field of herbalogy by Africans, yet most of this knowledge was passed down orally and yet to become part of the W3\estern pharmacopoeia.

****

THE VANISHING EVIDENCE OF CLASSICAL AFRICAN CIVILIZATIONS:
by Prof. Manu Ampim

The "Vanishing Evidence" series is a general summary of years of detailed observation and research. The full documentation supporting the conclusions expressed in this series of articles, including dozens of photographs, will be published in my forthcoming book, <u>Modern Fraud: The Forged Ancient Egyptian Statues of Ra-Hotep and Nofret</u>. I make no attempt here to "prove" such a complicated case as forgeries and the deliberate destruction of African artifacts in just a few short articles. Rather, this series is simply a preliminary report on my findings, which will be given extensive treatment in my book Modern Fraud. - MA

SCOPE OF "THE VANISHING EVIDENCE" SERIES

I originally wrote "The Vanishing Evidence of Classical African Civilizations" series in <u>The Gaither Reporter</u> in 1995-1996. The 3-part series documents the unintentional human-aided deterioration, as well as the deliberate and massive alteration, mutilation, and destruction of ancient Egyptian artifacts. The series covers three broad categories of *this vanishing evidence: the Temple Evidence (part I), the Tomb Evidence (part II), and the Museum Evidence (part III). Before writing the series, I had meticulously studied artifacts and images at temples, tombs, pyramids, museums, and ancient residential sites throughout Egypt and Nubia (1990, 1994-1995). I also carefully examined ancient Egyptian and Nubian artifacts at nearly all of the major museums, institutes, and libraries in eleven European countries (1989-1990), and in dozens of cities throughout the U.S. and Canada (1991-1992). Overall, I painstakingly studied well over a million images and artifacts before writing the "The Vanishing Evidence" series.*

My initial goal was simply to follow the path of research outlined by the eminent scholar, Dr. Chancellor Williams in <u>The Destruction of Black Civilization</u> *(1974), and conduct primary research on ancient Black civilizations. As Williams pointed out, the accurate knowledge of African science, social organization, and advanced spiritual systems can only be known from an unbiased first-hand study of the history and artifacts. Though I began my extensive multi-country research tour with a mission of documenting the specific details of the advanced African civilizations of ancient Egypt and Nubia, a quite different and unexpected theme emerged in every region that I studied. I was struck by the specific patterns of deliberate alterations and defacement of the artifacts. These patterns became clear after I had visited the initial group of European museums. What I noticed most was that these acts were more than simply knocking off the noses of statues, they involve a much broader assault on ancient Black images.*

The pair statue from the Tomb of Ikhetneb has lost most of its original paint on the faces and upper body of the two images, while much of the original paint on the legs and feet still survive. It seems evident that the darker paint has been deliberately erased on the face and upper body, thus giving the images the illusion of a white-skin appearance.

METHODS OF DESTRUCTION AND ALTERATION

My observations of the evidence are careful and are supported by photographic and video documentation. The extensive study of over a million artifacts and images gave me the keen power of observation to make analyses of the same materials that are also available to Egyptologists, although most of them have not studied the volume of the materials that I have examined. When the artifacts in question were examined meticulously, I found that the methods used to change or damage the images are varied, and they include: subtly altering the shape of the nose on statues by using some type of sanding device; adding on false noses; reshaping or completely destroying the face on temple and tomb walls; lightening the color of the face and body paint on statues and paintings; completely eliminating the paint on statues and paintings and thus making the images appear "white"; destroying the lower facial structure, particularly the chin and jaw area; putting in false bluish-gray inlaid eyes; plastering over temple images with White Portland Cement during ongoing "conservation" work; and creating outright forgeries! These acts of fraud and deliberate destruction is what I call the handiwork of modern conspirators. I was stunned by such an angry, vicious, and widespread attack against Black images by the enemies of Classical African Civilizations.

MODERN FORGERIES

One of the most absurd invention of the conspirators is the forged statues of Ra-Hotep and Nofret in the Cairo Museum. I present detailed and concrete evidence in my forthcoming book, <u>Modern Fraud: The Forged Ancient Egyptian Statues of Ra-Hotep and Nofret</u> *(also see part 3 of this series on the Museum Evidence) that there is a list of specific artistic rules which are consistently applied to statue after statue throughout the pyramid age.*[1] *Although some writers would attempt to argue to the contrary, the fact that there was a body of specific artistic rules is demonstrated by the tens of thousands of statues and paintings that are presented in exactly the same way across various sites and cites throughout all of the major periods: the old, middle, and new kingdoms. Thus, these shocking forgeries make no artistic sense as they violate a long list of clearly defined rules, according to the ancient Egyptian rule system. Further, these fabricated European-looking statues do not make any cultural or historical sense, and it is consistent that they were found (or rather created by modern hands) during the 19th century, which has been called by some historians as "The Great Age of Fakes." There is no doubt that the Ra-Hotep and Nofret forgeries stand completely outside of the ancient Egyptian artistic, cultural, and historical reality.*

One of the great forgeries of the 19th century which has already been proven is that of the famous Queen Tetisheri. The statue of "Queen Tetisheri" was purchased by the British Museum in 1890, and this fake piece (with its facial features resembling most Europeans) was paraded around the world until it was first suspected as a fraud in 1984. This fake Tetisheri statue was showcased in the British Museum special exhibit on forgeries in 1990, but not before this forgery fooled experts and deceived the world for 100 years![2]

This limestone statuette inscribed with the name of Queen Tetisheri was long regarded as an important piece for the study of ancient Egyptian sculpture of the late 17th and early 18th dynasties. Now, however, this world renown statuette is regarded as a modern forgery.

WHO ARE THE CONSPIRATORS?

Who are the conspirators responsible for these acts of destruction, alteration and invention is a basic question that must be answered. Brian Fagan in his book on The Rape of Egypt *(1975) documents a list of culprits: the* **Christian Copts** *who destroyed statues and monuments; the* **conquering Arabs** *who dismantled ancient buildings; and the 18th and 19th century* **European travelers, adventurers, and archeologists** *who were treasure hunters, plunderers, and looters. It was during the 19th century that a vast number of ancient Egyptian artifacts were excavated, but most of these discoveries were not adequately* documented, with original on-site photographs and detailed field reports, which are now standard procedures within the archeological field. The systematic documentation of archaeological excavations did not develop until many years after volumes of artifacts were already taken from their original African sites and eventually placed in museums. Indeed, it is a sketchy record of how most of these artifacts were discovered and eventually made it onto museum floors.

This brings us to the identification of a large group of conspirators, who are the **handlers of the excavated artifacts.** This group includes **excavators, antiquities dealers, museum curators and directors, and restorers and conservationists**. Somewhere between the original excavation of the artifacts, to their transport, sale and acquisition, storage, cleaning, conservation, and finally their display on the museum floor there has been a diligent effort at altering, reworking, and "touching up" the facial features of countless statues. The conspirators who perpetrate these acts are behind the scenes actors who use a hit and run strategy of defacement and alteration of the art. They hide their hands from public view, but the results of their fraudulent and destructive activity is plain to see by anyone who carefully examines the evidence, as their pattern of fraud is highly distinctive. The lack of original documentation of many excavated artifacts has made it easy for these conspirators to commit their fraud. It is well known that many great artifacts of ancient Egypt were destroyed by the hands of plunderers during the widespread looting and trading in antiquities that went on for centuries, but what is not well known is the full story of the artifacts that did survive the wave of plunderers and make it into modern museum collections. I show in Modern Fraud that a large number of these surviving artifacts have undergone a racial makeover at the hands of modern conspirators. The results of their fraudulent work are the countless altered artifacts, reworked pieces, fake genealogies, and a host of forged statues.

The most recent revelation of the racist fabrications by this group of **handlers**, or more specifically **Western museum directors**, is reported in the current issue of Archaeology Magazine 54 (September-October 2001, p. 27). This report is associated with an article by Peter Lacovara et al. Archaeology reported that in the absence of scholarship the directors of the Niagara Falls Museum in Ontario, Canada "fabricated pedigrees" for many of their Egyptian mummies in the mid-nineteenth century. The most imaginative of these fake pedigrees, or false identities, was created for a bearded male mummy of the Roman period. The museum officials invented the following *elaborate story for him which is a complete myth:* "General Ossipumphnoferu the General in Chief of Thotmes III. ... He was a man of military skill, also a famous magician. He was 60 years old when he died. The scar on his forehead was caused by an enraged elephant while defending the king from his onslaughts. A palace was erected for the general near that of the king." *The museum officials took their scandalous activity even further, as for many years the "general" was displayed in the coffin of Iawttayesheret, a high-ranking woman from the 25th dynasty, which was 700 years before his time! It is incredible that the directors of a public museum would take an unidentified Roman period mummy, with a European facial appearance, put him in a woman's coffin from 700 years earlier, and then create a bogus identity for him as a famous general during a period which was another 700 years earlier than the coffin he was buried in! Even though this mummy and other artifacts at the museum were not studied comprehensively by an Egyptologist, this is yet another case which documents that Western museum directors would go to any lengths in the 19th and early 20th century to falsify evidence.*

Currently, there is no doubt that this list of conspirators includes local **Egyptian government workers**, *who are carrying out many acts of destruction on a regular basis. These men either work for the Egyptian government on "conservation" projects, or for various European or North American archeological teams. On several occasions in the 1980s and 1990s, these unsupervised minimally-skilled government workers have been caught on video tape plastering over temple images and inscriptions! In fact, it is impossible to visit the Karnak Temple in Luxor and not see the recent defacement, and it is suspicious that with rare exception Egyptologists are silent about this matter.*

WHAT ARE THEIR MOTIVES

After examining a vast body of artifacts, it seems evident that the ultimate motives of these groups of conspirators from the 19th century to the present is to eliminate the Black images from the ancient Egyptian historical record. This motivation is consistent with the racist views of many of the 19th and early 20th century Egyptologists who made many ridiculous assertions about Black people having little to no role in ancient Egypt, and that this was a civilization founded by white or Semitic people from the North. These baseless claims were widespread within the ranks of Egyptologists, and they helped inspire both H. M. Herget's 1941 <u>National Geographic Magazine</u> *paintings of pale-skinned Egyptians and the imaginary white images created by*

Hollywood, which together have deceived the public for the past half century.[3] This nonsense was exposed in the 1950s by the late Senegalese scholar, Dr. Cheikh Anta Diop. Diop assembled an awesome body of evidence to document the Black foundation of ancient Egypt and to expose the dishonest discourse of Western Egyptologists who were, as he put it, "performing intellectual acrobatics" to avoid dealing with concrete evidence to support their contentions about the Northern origins of ancient Egyptian civilization.[4] The mainstream Egyptologist Bruce Trigger in <u>The American Discovery of Ancient Egypt</u> (1995) discusses Diop's impact and that it is because of his work that ancient Egypt is now seen by mainstream scholars as an African nation. Trigger also comments that "the white racist rhetoric that permeated most early twentieth-century writings about the development of Egyptian civilization has long been abandoned, [but] ideas formulated at the time have continued to influence thinking about the origins and nature of Egyptian civilization."[5]

The "white racist rhetoric" that Trigger describes as permeating early twentieth century writings is simply a continuation of the same racist views held in the 19th century, and it is within this climate that the behind the scenes handlers had both the motives and the opportunity to deface images, alter facial features, and create racist forgeries.

Notes

[1] I list some of these rules in Manu Ampim, "Ra-Hotep and Nofret: Modern Forgeries in the Cairo Museum?" in *Egypt: Child of Africa* (1994), pp. 207-212. I will discuss the full body of these rules in my forthcoming book, *Modern Fraud: The Forged Ancient Egyptian Statues of Ra-Hotep and Nofret.*

[2] For details see: Mark Jones, ed., <u>Fake: The Art of Deception</u> (1990), pp. 160, 162. The "Tetisheri" statue was first suspected to be a forgery in 1984 by Mr. W.V. Davies, Keeper of Egyptian Antiquities at the British Museum. See Davies, <u>British Museum Occasional Paper</u> (no. 36, 1984). For the significance of this forgery see Manu Ampim, "Ra-Hotep and Nofret: Modern Forgeries in the Cairo Museum?" in <u>Egypt: Child of Africa</u> (1994), pp. 207-208.

[3] William Hayes, "Daily Life in Ancient Egypt," <u>National Geographic Magazine</u> 80 (1941): 419-515. <u>H. M. Herget's 32 imaginary color paintings with pale-skinned Egyptians</u> are used as illustrations for Hayes' article, eventhough Hayes states "the Egyptians are, and always have been, Africans," and that they are a "brown" and "brunet" (i.e. dark brown or

reddish brown) skinned people. This deliberate public deceit and racism is still being carrying on by KMT Magazine, which uses these same dishonest images created by Herget more than a half century ago. For example see: KMT premiere issue (spring 1990), pp. 9, 11, 16; and KMT vol. 5, no. 4 (Winter 1994-95), pp. 45-46, 60. KMT now seems to have adopted a new favorite set of modern racist drawings by Winifred Brunton to illustrate its various articles; for the complete set of the Brunton drawings, see KMT, vol. 1, no. 4 (Winter 1990-91), pp. 52-61. Also see KMT, vol. 8, no. 1 (Spring 1997), p. 31; and vol. 12, no. 3 (Fall 2001), pp. 1, 74.

[4]Cheikh Anta Diop, The African Origin of Civilization: Myth or Reality (1974), trans. by Mercer Cook. The African scholars C.A. Diop and Theophile Obenga solidified their position of Egypt as a Black civilization at the historic 1974 "Peopling of Ancient Egypt" Symposium. See: UNESCO, The Peopling of Ancient Egypt and the Deciphering of the Meroitic Script: Proceedings of the Symposium held in Cairo, Egypt from 28 January to 3 February 1974 (The General History of Africa, Studies and Documents, no. 1, 1978), p. 102.

[5]Bruce Trigger, "Egyptology, Ancient Egypt, and the American Imagination" in The American Discovery of Ancient Egypt (1995), pp. 21-35.

[6]John Romer had already done thorough work in documenting the deterioration of the monuments, with his publication, Physical Deterioration of the Royal Tombs in the Valley of the Kings: A Progress Report on the 1977-1978 Season of the Brooklyn Museum Theban Expedition (1978). Along the same lines, Romer wrote The Rape of Tutankhamun (1993), which was later turned into a one-hour PBS television documentary in 1994.

Prof. Manu Ampim is a historian and primary researcher on African and African American culture & history. He is director of Advancing The Research

****

10 African Inventions That Changed the World

By Rising Africa, Jun 18, 2015

https://www.risingafrica.org/success-stories/education/10-african-inventions-that-changed-the-world/

Mathematics

The invention of mathematics is placed firmly in African prehistory. The oldest known possibly mathematical object is the Lebombo bone, which was discovered in the Lebombo Mountains of Swaziland and dated to approximately 35,000 B.C. Many of the math concepts that are learned in school today were also developed in Africa. Over 35,000 years ago, Ancient Egyptians scripted textbooks about math that included division and multiplication of fractions and geometric formulas to calculate the area and volume of shapes. MEDICINE: Many treatments used today in modern medicine were first employed in Africa centuries ago. The earliest known surgery was performed in Egypt around 2750 B.C. Medical procedures performed in ancient Africa before they were performed in Europe include vaccination, autopsy, limb traction and broken bone setting, brain surgery, skin grafting, filling of dental cavities, installation of false teeth, what is now known as Caesarean sections, anesthesia and tissue cauterization.

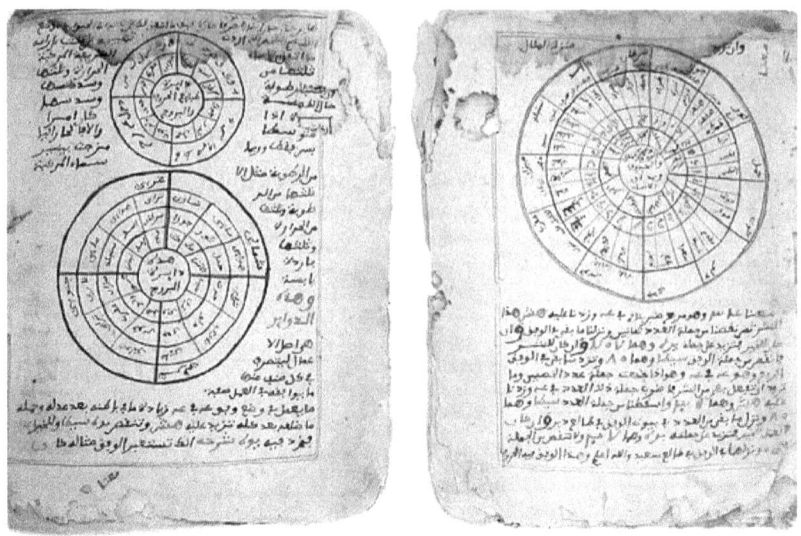

A famous example of a mathematical and astronomical manuscript from medieval Timbuktu

Medicine

Many treatments used today in modern medicine were first employed in Africa centuries ago. The earliest known surgery was performed in Egypt around 2750 B.C. Medical procedures performed in ancient Africa before they were performed in Europe include vaccination, autopsy, limb traction and broken bone setting, bullet removal, brain surgery, skin grafting, filling of dental cavities, installation of false teeth, what is now known as Caesarean sections, anesthesia and tissue cauterization.

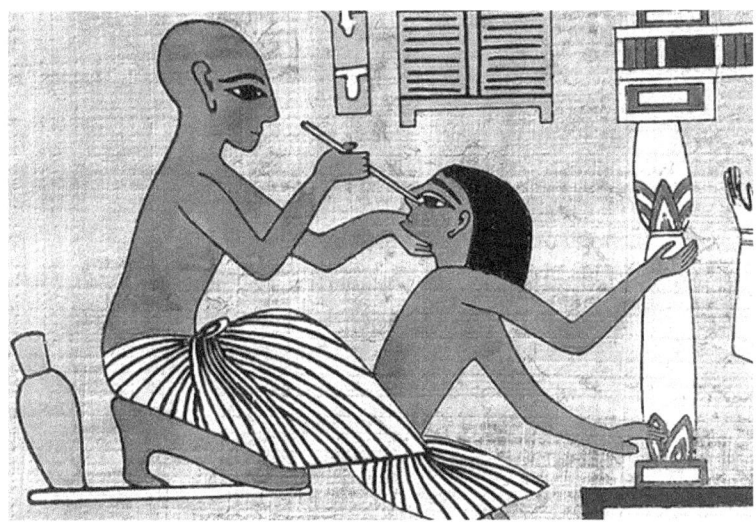

Architecture and Engineering

The African empire of Egypt developed a vast array of diverse structures and great architectural monuments along the Nile, among the largest and most famous of which are the Great Pyramid of Giza and the Great Sphinx of Giza. Later, in the 12th century there were hundreds of great cities in Zimbabwe and Mozambique made of massive stone complexes and huge castle-like compounds. In the 13th century, the empire of Mali boasted impressive cities, including Timbuktu, with grand palaces, mosques and universities.

Mining of Minerals

The oldest known mine on archaeological record is the "Lion Cave" in Swaziland, which radiocarbon dating shows to be about 43,000 years old. The ancient Egyptians mined a mineral called malachite. While the gold mines of Nubia were among the largest and most extensive in the world.

Metallurgy and Tools

Many advances in metallurgy and tool-making were made across the entirety of ancient Africa. These include steam engines, metal chisels and saws, copper and iron tools and weapons, nails, glue, carbon steel and bronze weapons and art. In places like Tanzania, Rwanda and Uganda, the advances in metallurgy and tool-making surpassed those in Europe.

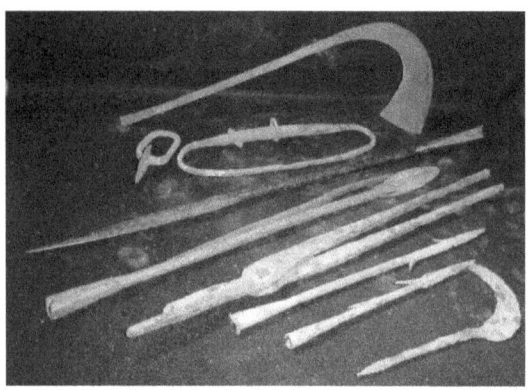

Navigation

Some people insist that Africans couldn't have made it to the New World first simply because they didn't have the skill and resources to sail across the Atlantic. As it turns out, that's completely false.

Historians have discovered evidence that suggests Africans were masters at building ships and that it was actually a part of their tradition. Shipbuilding and sailing are over 20,000 years old in the Sahara, and cave wall paintings of ancient ships were displayed in NationalGeographic magazine years ago. With those shipbuilding skills and the navigation skills that were noted by other historians of the time, the myth that Africans wouldn't have been able to sail to the New World becomes officially debunked. As Dr. Julian Whitewright, a maritime archaeologist at the University of Southampton, explained, the voyage from Africa on ancient ships was "quite a plausible undertaking, based on the capabilities of the vessel of the period and historical material stating it took place."

Evidence suggests that ancient Africans sailed to South America and Asia hundreds of years before the Europeans, debunking the propaganda that Europeans were the first to sail to the Americas. Many ancient societies in Africa built different types of boats, from small vessels to large ships that could carry up to 80 tons.

Astronomy

Several ancient African cultures birthed discoveries in astronomy. Many of these are foundations on which we still rely, and some were so advanced that their mode of discovery still cannot be understood. The Dogon people of Mali amassed a wealth of detailed astronomical observations. They knew of Saturn's rings, Jupiter's moons, the spiral structure of the Milky Way and the orbit of the Sirius star system. A structure known as the African Stonehenge in present-day Kenya (constructed around 300 B.C.) was a remarkably accurate calendar.

The Dancing Stones of Namoratunga, Kenya

This ancient site goes back over 2000 years, comprises of 19 basalt one-metre-high stones, almost perfectly cylindrical, some tilted over, others fallen completely, are arranged in rows forming a suggestive pattern.

Since the 70s, scientists have taken sightings on seven prominent Borana stars - Triangulum, Pleiades, Bellatrix, Aldebaran, Central Orion, Saiph and Sirius, as they would have appeared during this period. They have found that these stone pillars make 25 alignments with these seven stars.

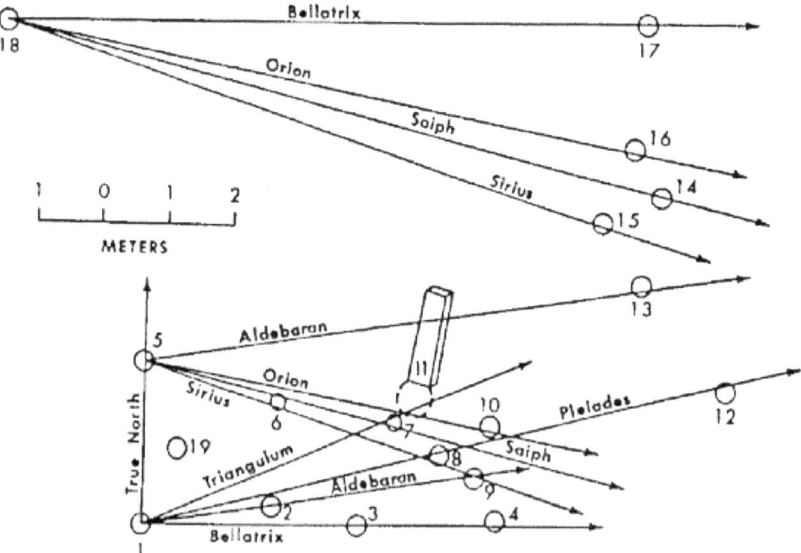

Stone alignments relative to the seven stars at Namoratunga II.

Almost two centuries before Roman emperor Julius Caeser commissioned a calendar to regular activities in his empire, Africans in this region made use of these pillar alignments with the constellations develop a sophisticated calendar of 12 months 354 days, like their neighbours the ancient Egyptians.

Philosophy

Philosophy in Africa has a long history dating from pre-dynastic Egypt and continuing through the birth of Christianity and Islam. One of the earliest works of political philosophy was the *Maxims of Ptah-Hotep*, which were taught to Egyptian schoolboys for centuries.

International Trade

Evidence shows that international trade was first developed between Africa and Asia, and among these international trade contacts were the exchange of ideas and cultural practices that laid the foundations of the earliest civilizations of the ancient world.

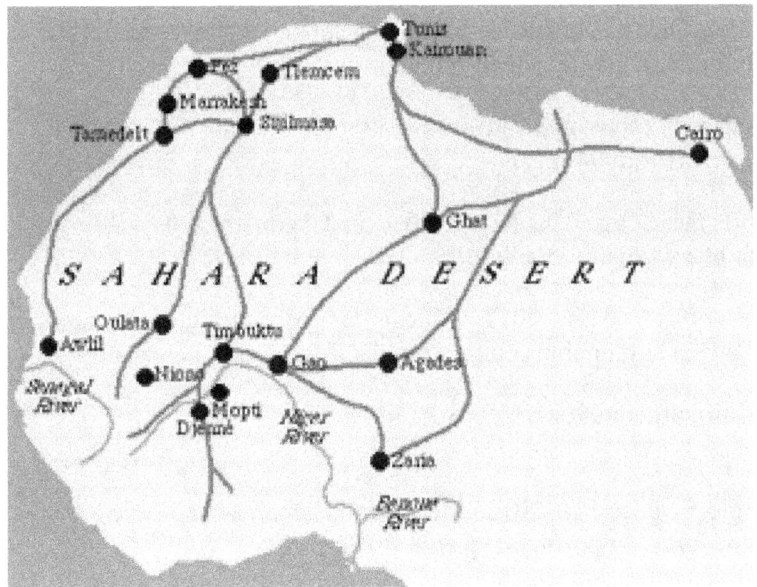

****

Sources:

West African Contributions to Science and Technology (Reklaw Education Lecture Series Book 11) Kindle Edition

- McLaren, W. August 2006. 'Mohammed Bah Abba And His Pot-in-Pot' http://tinyurl.com/986etkj

- Treacy, M. May 2012. 'Device Turns Your Sneaker into a Portable Cell Phone Charger'

  http://tinyurl.com/8jnwt8z

- Ashoka Innovators for the Public. 2008. 'Ashoka Innovators for the Public: Mohammed Bah Abba'

  http://tinyurl.com/9nlrwur

- Biofuels Digest. May 2012. 'CleanStar launches first cooking fuel facility in Mozambique; alternative to

  charcoal cooking, heath risks, at hand in Africa' http://tinyurl.com/94lmzal

- Noble Prize.Org. The Noble Prize in Chemistry 1999: Ahmed Zewail http://tinyurl.com/9jknsbr

- Novozymes. May 2012. 'CleanStar Mozambique launches world's first sustainable cooking fuel facility'

  http://tinyurl.com/9ozs7lu

- Google Science Fair 2012 'Meet our 15 finalists and Science in Action winner' http://tinyurl.com/8fjlw82

- RNW Africa Desk. February 2012. 'The Cardiopad: an African invention to save lives' http://tinyurl.com/9btaxfv

- The Right Livlihood Award. Aklilu Lemma (Ethiopia) http://tinyurl.com/9dd8j65

- Wikipedia. Sema Sgaier http://en.wikipedia.org/wiki/Sema_Sgaier

- Wikipedia. Raphael Armattoe http://en.wikipedia.org/wiki/Raphael_Armattoe

4. 4. Zaslavsky, C. "The Yoruba Number System." *Blacks in Science: Ancient and Modern.* 110 – 127 (1983).

5. 5. Lynch, B. M. & Robbins, L. H. Science 4343, 766 – 768 (1978).

6. 6. Adams, H. "African Observers of the Universe: The Sirius Question." *Blacks in Science: Ancient and Modern.* 27 – 46 (1983).

7. 7. Brooks, L. *African Achievements: Leaders, Civilizations and Cultures of Ancient Africa.* (1971).

8. 8. Shore, D. "Steel-Making in Ancient Africa." *Blacks in Science: Ancient and Modern.* 157 – 162 (1983).

9. 9. Asante, M. et al. "Great Zimbabwe: An Ancient African City-State." *Blacks in Science: Ancient and Modern.* 84 – 91 (1983).

Northern Africa

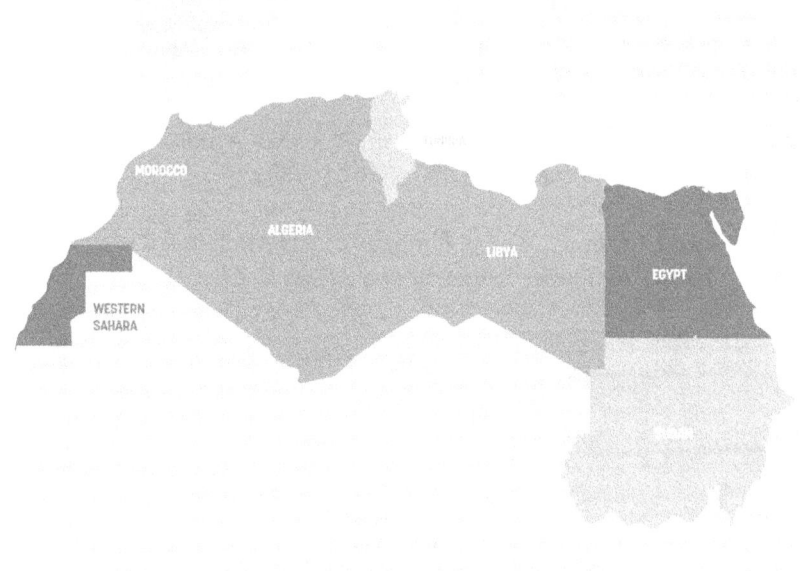

Before the Pharaohs
Egypt's Mysterious Prehistory

Edward F. Malkowski

This book explores the possibilities of cultures existing before and independent of the establishment of the Egyptian civilization.

"Geologists, archeologists, historians have been investigating prehistory for many years, which has resulted in volumes of analysis and opinion. Essentially, there are two kinds of evidence available: the physical, or archeological, and the historical (what the ancient people themselves reported about their history).

Physical evidence and historical 'facts' are often viewed with certain bias. This bias is a set of assumptions an individual brings to the evaluation of evidence.

There are different interpretations of history; researchers subject to different political and philosophical persuasions argue over interpretations of evidence. One version is common knowledge, taught to us from an early age, and we accept it as fact. Those responsible for our educational institutions consider other versions of interpreting information as maverick at best. As a result, researchers who are considered mavericks may be called irresponsible and speculative, and charged with misinterpreting the evidence. In a search for the truth, when significant portions of the story are missing, the interpretation is often determined by the interpreter's underlying philosophy."

****

Ancient Egyptian Inventions
by Mark Millmore
https://discoveringegypt.com/ancient-egyptian-inventions/

The Egyptian's inventions were many and it might be easier to list the things they did not invent such as the wheel; not unexpected in a country where everyone travels on water.

Most scholars now believe that isolated civilizations first arose independently at several locations; initially in Mesopotamia around Tigris and Euphrates rivers and, a little later, in Egypt and the eastern Mediterranean. Other civilizations arose in Asia along the Indus River in modern India and the Yellow River in what is now China.

All these early civilisations had to invent or discover everything for themselves because unlike later civilisations such as the Greeks in the west or the Chinese in the east, they had no one to learn from. Therefore, the Egyptians had to invented mathematics, geometry, surveying, metallurgy, astronomy, accounting, writing, paper, medicine, the ramp, the lever, the plough, mills for grinding grain and all the paraphernalia that goes with large organised societies.

So how do we define Egyptian inventions today? It is very difficult to determine because three thousand years is a long time for discoveries to be made and lost or appropriated by others. For example the Greeks sometimes take the credit for inventing mathematics but they learned their math from the Egyptians then later developed and improved upon what the Egyptian achieved.

3000 BC appears to have been a critical time for the development of technology, especially metal making. The Egyptians as well as the Mesopotamians independently discovered that by mixing a small quantity of tin ore with copper ores they could make bronze which is harder and more durable. This set off a chain of connected innovations that could not have happened without the primary discovery
.

The Pyramids

The Step Pyramid of Djoser is the oldest pyramid in Egypt. It was built about 4,700 years ago.

The oldest pyramid was erected for King Zoser between 2667-2648 BC. In fact it is the first monumental stone building designed and constructed that we know of.

Writing

Along with the Mesopotamians, the Egyptians were the first people to develop their language into a codified form of writing. All early forms of writing were pictograms – pictures. All writing systems developed in this way but their original forms become lost as the pictures are refined into abstract forms. What is interesting about the Egyptians is that although their writing changed to the abstract form of Hieratic they deliberately preserved the hieroglyphic pictures in their original forms.

Papyrus Sheets

Papyrus sheets are the earliest paper-like material – all other civilizations used stone, clay tablets, animal hide, wood materials or wax as a writing surface. Papyrus was, for over 3000 years, the most important writing material in the ancient world. It was exported all around the Mediterranean and was widely used in the Roman Empire as well as the Byzantine Empire. Its use continued in Europe until the seventh century AD, when an embargo on exporting it forced the Europeans to use parchment.

Black Ink

The Egyptians mixed vegetable gum, soot and bee wax to make black ink. They replaced soot with other materials such as ochre to make various colours.

The Ox-drawn Plough

Using the power of oxen to pull the plough revolutionised agriculture and modified versions of this Egyptian invention are still used by farmers in developing countries around the world.

The Sickle

The sickle is a curved blade used for cutting and harvesting grain, such as wheat and barley.

Irrigation

The Egyptians constructed canals and irrigation ditches to harness Nile river's yearly flood and bring water to distant fields.

Shadoof

The Shadoof is a long balancing pole with a weight on one end and a bucket on the other. The bucket is filled with water and easily raised then emptied onto higher ground.

The Calendar

The Egyptians devised the solar calendar by recording the yearly reappearance of Sirius (the Dog Star) in the eastern sky. It was a fixed point which coincided with the yearly flooding of the Nile. Their calendar had 365 days and 12 months with 30 days in each month and an additional five festival days at the end of the year. However, they did not account for the additional fraction of a day and their calendar gradually became incorrect. Eventually Ptolemy III added one day to the 365 days every four years.

Clocks

In order to tell the time Egyptians invented two types of clock. Obelisks were used as sun clocks by noting how its shadow moved around its surface throughout the day. From the use of obelisks they identified the longest and shortest days of the year. An inscription in the tomb of the court official Amenemhet dating to the 16th century BC shows a water clock made from a stone vessel with a tiny hole at the bottom which allowed water to dripped at a constant rate. The passage of hours could be measured from marks spaced at different levels. The priest at Karnak temple used a similar instrument at night to determine the correct hour to perform religious rites.

The Police

During the Old and Middle Kingdoms order was kept by local officials with their own private police forces. During the New Kingdom a more centralized police force developed, made up primarily of Egypt's Nubian allies, the Medjay. They were armed with staffs and used dogs. Neither rich nor poor citizens were above the law and punishments ranged from confiscation of property, beating and mutilation (including the cutting off of ears and noses) to death without a proper burial. The Egyptians believed that a proper burial was essential for entering the afterlife, so the threat of this last punishment was a real deterrent, and most crime was of a petty nature.

"They went to the granary, stole three great loaves and eight sabu-cakes of Rohusu berries. They drew a bottle of beer which was cooling in water, while I was staying in my father's room. My Lord, let whatsoever has been stolen be given back to me." (Eighteenth Dynasty)

Surgical Instruments

The Edwin Smith Papyrus shows the Egyptians invented medical surgery. It describes 48 surgical cases of injures of the head, neck, shoulders, breast and chest. It includes a list of instruments used during surgeries with instructions for the suturing of wounds using a needle and thread. This list includes lint, swabs, bandage, adhesive plaster, surgical stitches and cauterization. It is also the earliest document to make a study of the brain. The Cairo Museum has a collection of surgical instruments which include scalpels, scissors, copper needles, forceps, spoons, lancets, hooks, probes and pincers.

Wigs

During the hot summers many Egyptians shaved their heads to keep them clean and prevent pests such as lice. Although priests remained bald as part of their purification rituals, those that could afford it had wigs made in various styles and set with perfumed beeswax.

Cosmetic Makeup

The Egyptian invented eye makeup as far back as 4000 B.C. They combined soot with a lead mineral called galena to create a black ointment known as kohl. They also made green eye makeup by combining malachite with galena to tint the ointment. Both men and women wore eye makup; believing it could cure eye diseases and keep them from falling victim to the evil eye.

Toothpaste

At the 2003 dental conference in Vienna, dentists sampled a replication of ancient Egyptian toothpaste. Its ingredients included powdered of ox hooves, ashes, burnt eggshells and pumice. Another toothpaste recipe and a how-to-brush guide was written on a papyrus from the fourth century AD describes how to mix precise amounts of rock salt, mint, dried iris flower and grains of pepper, to form a "powder for white and perfect teeth."

Mummification

The Egyptians were so expert at preserving the bodies of the dead that after thousands of years we know of the diseases they suffered such as arthritis, tuberculosis of the bone, gout, tooth decay, bladder stones, and gallstones; there is evidence, too, of the disease bilharziasis (schistosomiasis), caused by small, parasitic flatworms, which still exists in Egypt today. There seems to have been no syphilis or rickets.

****

The father of medicine is Imphotep, who lived around 2655-2600 BC, not the Greek Hippocrates who lived around 460 BC – 370 BC). Imhotep (called Imuthes (Ιμυθες) by the Greeks), was an Egyptian polymath (a person whose expertise spans a significant number of different subject areas), who served under the Third Dynasty King, Djoser, as Chancellor to the Pharaoh and High Priest of the Sun God Ra at Heliopolis. He is considered to be the first architect engineer and physician in early history.

****

Western Africa

Ancient West Africa (2,000 BC – 500 AD)

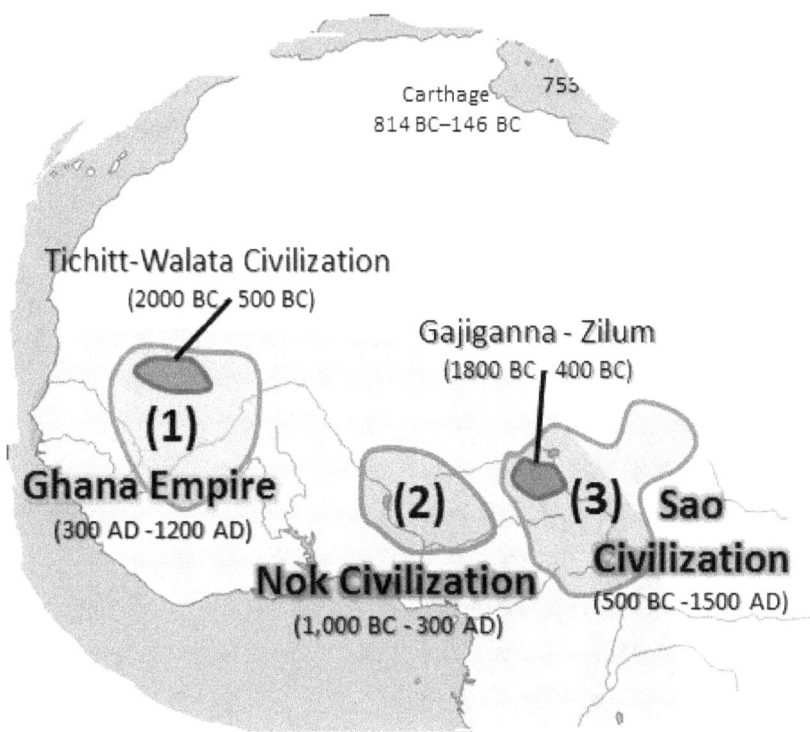

****

Nok Society

Modern Region	Past Civilization	Item		Date	Information Source
Nigeria	Nok	Advance Smelting	Iron	900 BC	Discovered in 1928, *When We Ruled*, by Robin Walker
Nigeria	Igbo-Ukwu	Advance and Objects	Copper Bronze	900 AD	Discovered in 1938, *When We Ruled*, by Robin Walker

from Wikipedia:

The **Nok culture** is an early Iron Age population whose material remains are named after the Ham village of Nok in Kaduna State of Nigeria, where their terracotta sculptures were first discovered in 1928. The Nok Culture appeared in northern Nigeria around 1500 BC [1] and vanished under unknown circumstances around 500 AD, thus having lasted approximately 2,000 years.

Iron use, in smelting and forging tools, appears in Nok culture by at least 550 BC and possibly a few centuries earlier. Data from historical linguistics suggest that iron smelting was independently discovered in the region by 1000 BC. Scientific field work began in 2005 to systematically investigate Nok archaeological sites, and to better understand Nok terracotta sculptures within their Iron Age archaeological context.

Nok sculpture, terracotta, Louvre

Emma George Ross
Department of Arts of Africa, Oceania, and the Americas, The Metropolitan Museum of Art:

Iron smelting and forging technologies may have existed in West Africa among the Nok culture of Nigeria as early as the sixth century B.C. In the period from 1400 to 1600, iron technology appears to have been one of a series of fundamental social assets that facilitated the growth of significant centralized kingdoms in the western Sudan and along the Guinea coast of West Africa. The fabrication of iron tools and weapons allowed for the kind of extensive systematized agriculture, efficient hunting, and successful warfare necessary to sustain large urban centers.

****

Dogon Society

Sirius in art and astronomy of Dogon

from Efstratios Theodossiou, Associate professor- University of Athens

Sirius B is a white dwarf discovered by Alvan G. Clark only in 1862 with the help of powerful telescope. The performance of Sirius B is kept in the secret rituals of an isolated and primitive tribe. It is sacred taboo of the race, so they did not call him by its name, but as Kize Uzi, meaning "little thing". Shown in the sacred architecture, carved in woodcut artwork and woven textiles in the race. Also this tribe, the Dogons, reaffirms its unity every 60 years with the Sigi celebration and constructing a large wooden mask for each separate community.

The Dogon tribe, in Mali of West Africa, lives near Tombouctou, in the highlands between Bantiagkara and Ompori. For Dogons, the main information about this tribe comes from Marcel Griaule and Germaine Dieterlen, two famous French anthropologists, who since 1930 and for about 20 years studying the habits and the customs of the Dogon tribe.

After so many years near Dogon in Mali, the two French anthropologists joined the group and the Dogons decided to reveal the depths of the worldview of their people. Two priests, the Yébéné (priest of Binou Yébéné) and Manda D'Orosongo (priest of Binou Manta), a chieftain, the Ongkoulou Ndola and Innekouzou Dolo priestess (Ammagiana) of Amma, the god-creator of the Universe, reported in Griaule, in 1946, that a dominant position in the mythology and cosmological legends of the tribe occupies Sirius, the brightest star of the night sky.

Then they drew with a stick on the ground the orbit of Sirius and its secondary star (Sirius B), saying that this bright is actually a double star system.

They argued, then, that the visible Sirius, calling the Sigi Tolo star, is located at the one end of the path of a smaller star, which they call Po Tolo or Yurugu Tolo (star of Yurugu). Yurugu means the Pale Fox, for this reason the work of Griaule and Dieterlen called "Le Renard Pâle", which in French means the Pale Fox. Yurugu, according to the beliefs of the tribe, was born as a being who is stigmatized by fate to endlessly chasing his female soul, which is his ideal duo. Originally tried to catch it, he has stolen from his mother Earth, a piece of cake -which came after his birth- thinking that it was the twin soul.

The stellar system of Sirius (Sigi Tolo)
A= Sirius, B = Sirius B (Po Tolo), Γ=Sirius Γ (Emme)
Sirius A and its invisible companion, Sirius B

As we can see from the figure of the motions, the Dogon placed Sirius at the focus of an ellipse and not in the center of a circle. This shortage represents the trajectory of the Po Tolo, the invisible companion of Sirius, which has a period of 50 years. According to Dogon, the star of Yurugu, the smallest and heaviest star in the sky is the beginning and end of all things. It consists of a metal named 'sagala', lighter than iron, but the star is so heavy that not getting all the power of the Earth to lift it. They also say that the star weighs as all seeds or all of the iron of the Earth.

All the above, although it is inexplicable how the Dogon people knew them, are true if this star, or like today called Sirius B, is a white dwarf, discovered by Alvan G. Clark only in 1862 with the help of powerful telescope. The performance of Sirius B is kept in the secret rituals of this isolated tribe. It is sacred taboo of the race, so they did not call him by its name, but as Kize Uzi, meaning the 'little thing'. Also Dogon tribe reaffirms its unity every 60 years with the Sigi celebration, constructing a large wooden mask for each separate community.

Dogon people say that in this system belongs and a third star, Sirius C, which they call Emme. Sirius C, which is greater than Sirius B, considered four times lighter and Dogons believe that moving to a higher orbit, and a period as the Sirius B, equal to 50 years. The respective positions of the two stars are such that the angle formed by the rays being correct. For Sirius C, the Dogons say that is the Sun of the Women (Nyan Nai) or a small Sun (Nai Dagi), which accompanies a planet, called Emme gia, or the driver of Emme or the Star of the Women (Nyan Tolo) or the Guide of goats (Enegirin). The female star moves on an ellipse with Sirius C in on focus of the elliptical orbit.

Central Africa

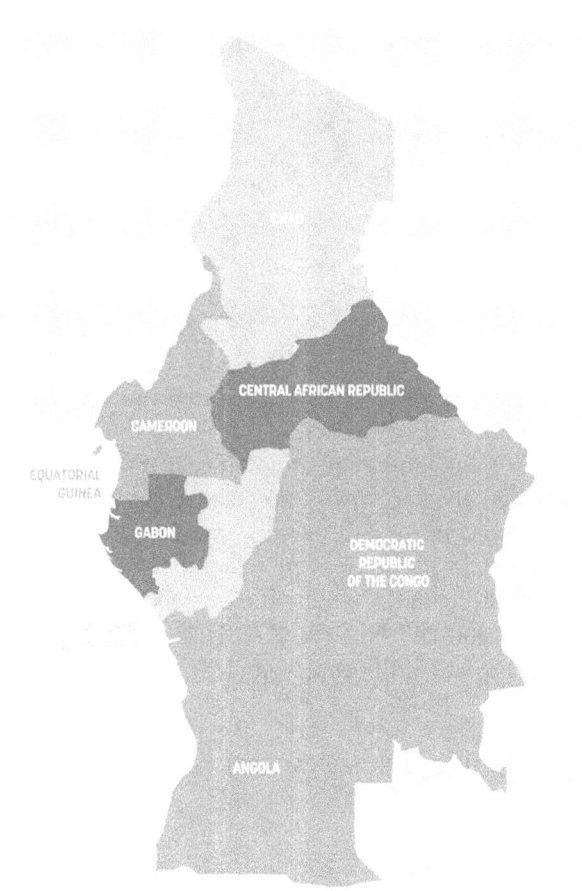

Sao Civilization

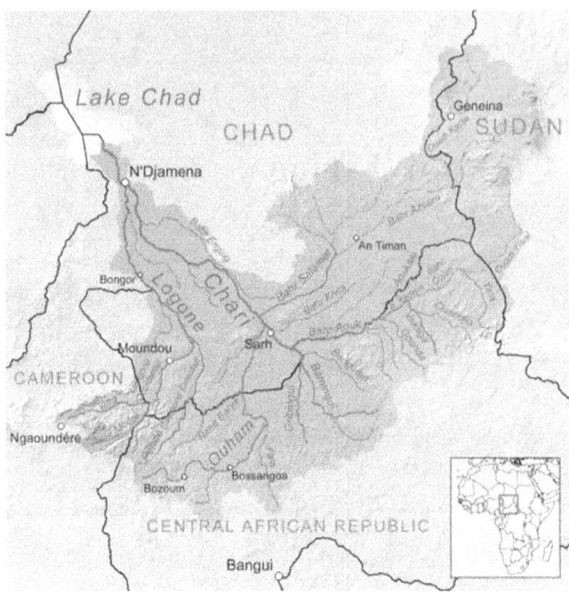

From archaeological evidence it is known that humans have inhabited Cameroon for at least 50,000 years, and there is strong evidence of the existence of important kingdoms and states in more recent times. Of these, the most widely known is Sao, which arose in the vicinity of Lake Chad, probably in the 5th century ce. This kingdom reached its height from the 9th to the 15th century, after which it was conquered and destroyed by the Kotoko state, which extended over large portions of northern Cameroon and Nigeria. Kotoko was incorporated into the Bornu empire in the late 19th century, and its people became Muslims.

The jewellery of the Saoian epoc, consisted of large and small beads strung together, which we can conjecture, became more sophisticated as they evolved.

BRONZE AGE: The evidence for this is found in the *cire perdu* lost wax process, [ca.3000 BC] which the Sao used to make bronze jewellery cast according to the ancient abovementioned process. Also discovered were bronze ritual cups, pendants, ankle- bracelets and weapons, all now with Europes private collectors instead of in our museums. The weaponry informs us that the Sao hunted and were well aware of the need to defend themselves against aggression. There is also evidence that they were fishermen.

REFERENCES:

1. Art and Mankind; Larousse Enc of Prehistoric and Ancient Art, [1962].

2. Lebeuf, Jean-Paul, and Annie Masson Detourbet (1950). La civilization du Tchad, Paris

****

Eastern Africa

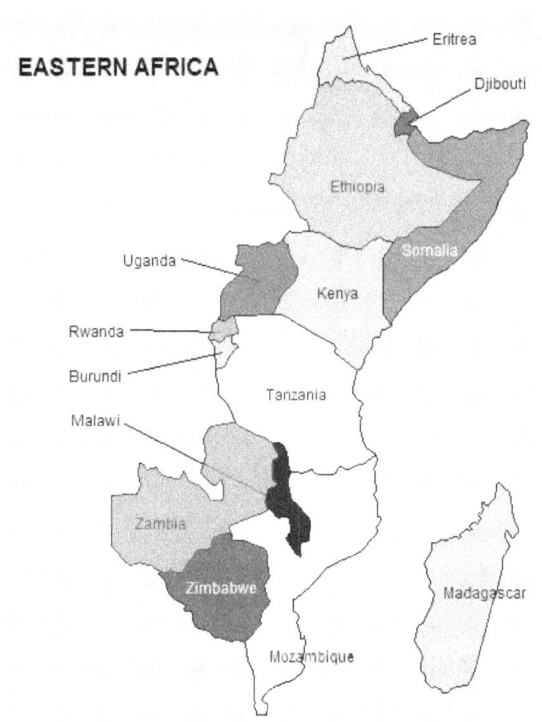

Kingdom of Kush, Rival to Egypt

The Kushite imperial capital was located at Meroe. In early Greek geography, the Meroitic kingdom was known as Aethiopia. The Kingdom of Kush with its capital at Meroe persisted until the 4th century AD

The pyramids of Meroe – UNESCO World Heritage

By **Núria Castellano**

Rising high in the sky in the modern-day nation of Sudan is a plethora of pyramids. They mark the site of the ancient city. The center of a powerful civilization, Meroë served as the capital city of Kush whose robust culture thrived for centuries. Their grand architecture and works of art left a lasting testament to the greatness of the Nubian kings and queens.

Located in the desert sands near the Nile in modern Sudan, the ancient culture of Nubia played a decisive role in shaping Egypt from the eighth century B.C., serving as that kingdom's 25th dynasty in the Late Period. After the Nubian pharaohs lost power, they retreated south from Egypt to form the Kingdom of Kush, which thrived in splendid isolation as the rest of Egypt suffered through repeated invasions from Assyrians, Persians, and Greeks.

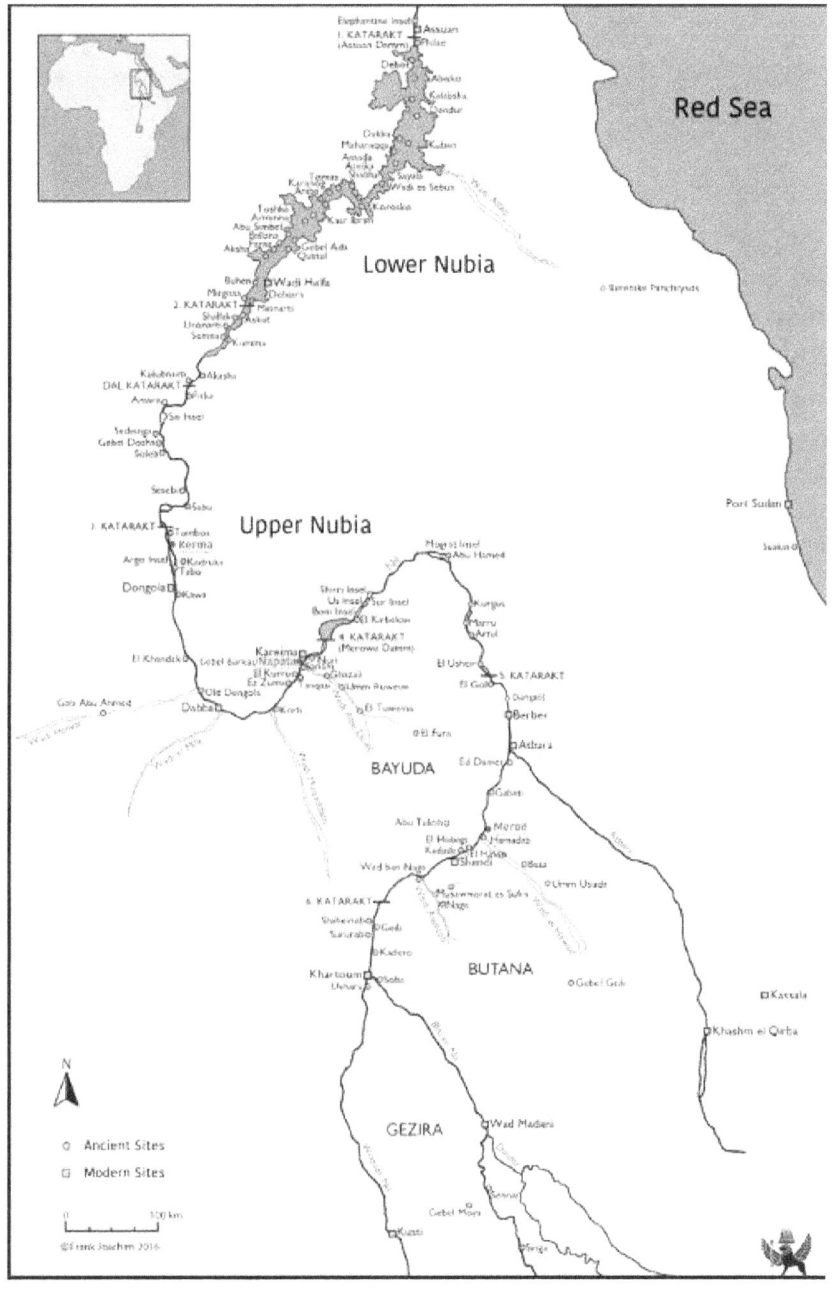

A map of ancient Kush: Lower and Upper Nubia, Bayuda and Butana, in modern day Sudan. The core territories of Kush stretched more than a thousand km, from the first cataract around Aswan (the traditional frontier), to the southern Butana steppe. More than 80 ancient sites can be seen, most of which have occupation dates from the Napatan and Meroitic periods of Kush. These are mostly towns, cities and temple complexes, which often have earlier, and later occupation dates as well. The three successive capital cities of Kerma, Napata and Meroë are highlighted in red. Other important sites include: Qasr Ibrim, Karanog, Amara East, Sedeinga, Tabo, Kawa, El Kurru, Sanam, Nuri, Dangeil, Homat el Hamadab, El Hassa, Basa, Wad Ben Naqa, Naqa and Musawwarat es Sufra.

Sebiumeker, the Meriotic god of procreation, was associated with the Egyptian creator god, Atum. Sandstone statue from Meroë, first century B.C. Carlsberg Glyptotek, Copenhagen Photograph by Prisma/Album

Because of Meroë's distance, the Kushites were able to retain their independence, developing their own vibrant hybrid of Egyptian culture and religion until well into the fourth century A.D. With access to mines and minerals, the Meroites were expert goldworkers. They built temples, palaces, and royal baths in their capital. Perhaps their grandest achievements are the more than 200 pyramids built at the necropolis at Meroë, giving Sudan more pyramids than all of Egypt. Tall, slender, graceful: These monuments bear witness to the lasting splendor that was Kush.

Egypt's 25th dynasty

In the late 20th century A.D. Swiss archaeologist Charles Bonnet spent decades excavating the lands surrounding the southern Nile. He found evidence of a civilization grown rich from trade and abundant with fields and livestock, a kingdom distinct from Egypt with its own material culture and traditions. This civilization grew in power just as Egypt's Middle Kingdom was in decline around 1785 B.C. By 1500 B.C., the Nubian empire roughly stretched from Wadi Halfa south to Meroë.

Centered on its original capital at Napata, the Nubian ruling dynasty continued to flourish militarily and economically through the ninth century B.C. Around 730 B.C., the Nubian king, Piye, successfully invaded and conquered Egypt, extending his control to the whole Nile Valley. Piye became the first pharaoh of Egypt's 25th dynasty (ca 770-656 B.C.), the so-called Black Pharaohs.

Meroë's Powerful Sisters

The tomb of Queen Khennuwa in Meroë. The art adorning this fourth-century B.C. tomb is related to the style of Egypt's 25th dynasty, the Black Pharaohs, four centuries earlier.

Photograph by Pawel Wolf, German Archaeological Institute

Ono of the most remarkable features of Meroitic civilization was its strong queens. In his *Geography,* Greek historian Strabo wrote of a queen called "Candace" who signed a peace treaty with the emperor Augustus. Candace, in fact, means "sister," and was the title given to Kushite queens . There were many queens in Meroë , such as Amanirenas—the "Candace" Strabo was referring to—and her successor, Amanishakheto, whose treasure was looted in 1834. Archaeologists have recently been studying the funerary chamber of another queen, Khennuwa, whose tomb was excavated by George Reisner in 1922.

Piye died in 715 B.C., having reigned 35 years. Although he had returned to Nubia after conquering Egypt, he wished to be buried in the Egyptian style, a request his subjects granted. Entombed in a pyramid, Piye was the first pharaoh in more than 500 years to be buried this way.

The 25th dynasty would last for three-quarters of a century. Its reign ended in turmoil when an Assyrian invasion of Egypt caused it to fall from power. The victors struck the names of the 25th dynasty from monuments across Egypt, destroying their statues and stelae to erase their names from history.

A small chapel tomb (ca 13th century B.C.) at the necropolis of the ancient city of Deir el Medina, near Luxor in Egypt. The pyramids of Meroë adopted a strikingly similar design . Photograph by Wael Hamdan, Age Fotostock

After the defeat, the Nubians retreated to Napata, only to be forced farther south at the beginning of the sixth century B.C., when Pharaoh Psamtek II, part of the 26th dynasty, sacked Napata. The Kushites designated the city of Meroë, which sat farther south along the Nile, as the new capital. This new location was carefully considered. Not only strategically positioned at the crossroads of inland African trade routes and caravan trails from the Red Sea, the land around Meroë was also fertile and blessed with significant natural resources—iron and gold mines that fostered the development of a metals industry, especially gold working.

Royal Tombs

Gold ring found in a Meroë tomb. This Egyptian-influenced piece depicts an *udjat*—eye of Horus—flanked by two cobras. State Museum of Egyptian Art, Munich Photograph by BPK/Scala, Florence

The Kushites' burial culture had been touched by a synthesis of Egyptian and African religious and cultural practices. Even after relocating south, the Kushite kings continued to be buried in the necropolis at Nuri, near Napata, a center of the cult of the Egyptian god Amun.

Meroë would become the preferred necropolis later, around 250 B.C. There are two main burial areas: the south cemetery and the northern burial ground. The south cemetery was the oldest. When it reached capacity, the northern burial ground was begun . The northern area today contains the best preserved of the pyramids at Meroë . Some of the most impressive tombs here are the final resting places for 30 kings, eight queens, and three princes.

Meroë's earliest pyramids were step pyramids. Scholars have speculated that cylinders or spheres may have once topped the pyramids, made of materials that have since been destroyed or perished. The later structures, built in the third century A.D., are simpler with smooth, steep sides. In spite of the clear influence of the classic Egyptian design, Meroë's pyramids are notably smaller and generally lack the *pyramidion,* a pointed capstone. Their design more closely resembles the chapel pyramids built at Deir el Medina near Luxor. These were built during Egypt's New Kingdom period (1539-1075 B.C.), a period when many Egyptian customs began to appear in Kushite culture.

The stones were set in place with a *shaduf,* or shaft, a device used as a lever to raise stone blocks. The outside was faced with brick and then covered with brightly painted plaster.

There are 41 tombs in Meroë's north cemetery , 38 of which belong to monarchs who ruled the region between 250 B.C. and A.D. 320. Photograph by Fabian Von Poser, Age Fotostock

Steps were carved into the rock to the east of each pyramid leading down to a sealed entrance. Behind it lay underground rooms with vaulted ceilings: three for a king and two for a queen. In the oldest pyramids, the burial chamber was decorated with scenes from the Egyptian Book of the Dead. A wooden coffin, depicting the dead person's face, was placed in the burial chamber. The sacrificed bodies of animals and, in some cases, of human servants were placed nearby.

Attached to one side of a standard Meroë pyramid was a chapel, its entrance formed by twin tapering pylons. Inside, it was common to place a stela, an offering table, and a distinctive element of Meroë culture: a statue of the *ba*—the aspect of the human soul believed to give the deceased their individuality—depicted as the body of a bird and a human head.

The endurance of Kush

Kush prospered for centuries, but Queen Cleopatra's death in 30 B.C. brought change. Egypt became a province of the nascent Roman Empire, straining the fragile truce that the Kushites had brokered with Rome. Tax revolts in Upper Egypt led to Roman incursions into Kushite territory, threatening their lucrative gold mines. Meroite forces attacked Roman troops in Aswan—the most southerly frontier of the Roman world—led by the fearsome Queen Amanirenas of Meroë. In his great work *Geography,* the Greek scholar Strabo describes her as Queen Candace, "a masculine woman ... who had lost an eye." This memorable commander was eventually beaten back to Meroë, but from then on, the Meroitic civilization was largely left in peace.

A QUEEN'S TREASURES

The jewels of Meroë's first-century B.C. queen Amanishakheto were stolen in 1834 by Guiseppe Ferlini. Today they are on display in the State Museum of Egyptian Art, Munich, and the Egyptian Museum of Berlin.This gold ring is adorned with the head of the god Amun in the form of a ram.

Multicolored glass paste pieces inlaid with gold. The clasp depicts the god Amun with a ram's head before a chapel door.Photograph by Egyptian Museum of Berlin/BPK/Scala, Florence

This gold ring shows two royal figures holding a child. One adult may be Amanishakheto. The child may be her son and heir.Photograph by Egyptian Museum of Berlin/BPK/Scala, Florence

Meroë was abandoned in the fourth century A.D. Over the centuries, rumors spread of its monuments and the gold they contained, eventually reaching the Italian tomb robber Giuseppe Ferlini. In 1834 Ferlini arrived in Meroë, where he set about looting the graves. The damage Ferlini caused is still lamented by archaeologists, but the few exquisite artifacts he brought back opened the eyes of European scholars to this mysterious culture that had absorbed the ancient traditions of pharaonic Egypt.

Timeline: Pyramids and Power

A Meroë king standing before the Egyptian god Re. Engraved gold plaque, Archaeological Museum, Khartoum, Sudan

Photograph by Werner Forman/GTRES

8th Century B.C.
Construction of the south and west cemeteries begins in Meroë, then the second city of the Kingdom of Kush, whose capital was at Napata.

3rd Century B.C.
As space in Meroë's south cemetery runs out , expansion begins in the north cemetery of the city's growing necropolis.

250 B.C.
King Arakamani relocates the royal necropolis from near Napata to Meroë, which becomes the kingdom's spiritual and political capital.

1st Century A.D.
Queen Amanirenas leads her troops against the Romans. Her successor, Amanishakheto, is buried with costly grave goods.

2nd Century A.D.
Building methods change . The Meroë pyramids are faced with brick instead of stone, and then a layer of plaster, which is painted.

A.D. 350
An invasion by the kingdom of Axum brings Meroë 's dominance to an end. The city and royal necropolis are eventually abandoned.

A specialist in the later Egyptian dynasties, Núria Castellano is an Egyptologist at the University of Murcia, Spain.

Two of the most important resources of Ancient Kush were gold and iron. Gold helped Kush to become wealthy as it could be traded to the Egyptians and other nearby nations. Iron was the most important metal of the age. It was used to make the strongest tools and weapons.

****

Southern Africa

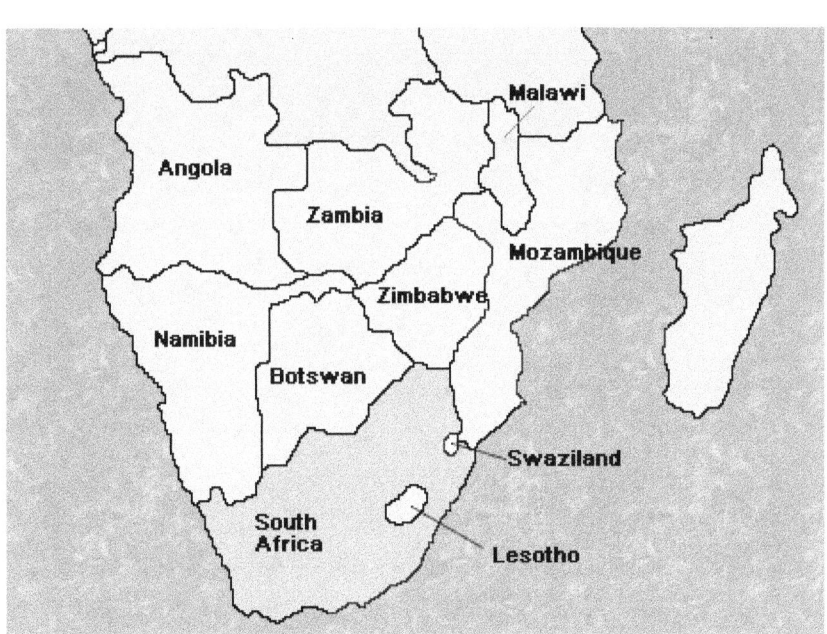

Hematite mining is conducted in the Ngwenya mountain area of Swaziland (41,200 BC)

Manganese mining is conducted at Chowa in Zambia (26,000 BC).

Black Contributions to Asia

5 Ancient Black Civilizations That Were Not in Africa

By K. Abel - April 16, 2014

The Minoans Ancient Greece

Archaeologist Manfred Bietak conducted extensive research on ancient Greek civilizations and their connections to ancient Egypt. Bietak uncarthed evidence from artwork as early as 7000

B.C. that depicts the early people inhabiting Greece were of African descent.

The Minoan culture of Ancient Greece reached its peak at about 1600 B.C. They were known for their vibrant cities, opulent palaces and established trade connections. Minoan artwork is recognized as a major era of visual achievement in art history. Pottery, sculptures and frescoes from the Minoan Bronze age grace museum displays all over the world. Palace ruins indicate remnants of paved roads and piped water systems.

Indus Kush Civilization

On March 3, 2000, historian Runoko Rashidi gave a lecture in Honolulu, Hawaii, about the presence of Black people in ancient and modern India. He stated that the face of India changed around 2000 B.C. when nomadic people Indo-Europeans or Aryans traveled to the Innis Valley and other fertile locations in southern India.

Prior to the invasion, Blacks in India built rich and advanced civilizations. Author Wayne Chandler recanted his amazing discoveries about Blacks in ancient India in his book "African Presence in Early Asia." The remarkable cities of Harrappa and Mohenjo-daro are only two of the many cities built by Black people. These cities cover large regions of northern India and modern-day Pakistan.

Ancient Mexico

The Olmecs were an ancient civilization in the Americas. Researchers such as Rashidi, Ivan Van Sertima and Alexander Von Wuthenau have discovered and shared evidence showing that the original inhabitants of Mexico were of African descent. The Olmecs were no different from people found in the Mende regions of West Africa.

Best known for carving the colossal stone heads that date back to 1100 B.C., more evidence of their existence before European explorers has been found. The Olmecs built pyramid-like structures made of mud in Mexico. They were also very artistic and created terracotta art that displayed common activities like pottery-making and wrestling. To add to their achievements, the Olmec people developed a calendar system around 3100 B.C.

Shang Dynasty of Ancient China

In a genetic study published in the *"Proceedings of the National Academy of Science Genetic,"* researchers found evidence showing the first African arrived in China about 60,000 years ago. Researcher and population geneticist Li Jin states, "Our work shows that modern humans first came to southeast Asia and then moved later to northern China. This supports the idea that modern humans originated in Africa."

A 2009 published essay from the *"Light Words from the Dark Continent; A Collection of Essays,"* by Nibs Ra and Manu Amun, offers insight to early Chinese civilizations. It states that the first documented governance in China was headed by the Shang or Chiang dynasty in 1500-1000 B.C. King T'ang or Ta, founder of the Shang dynasty, was of African descent. The Shang were also called Nakhi, which literally means "Black" (Na) and "Man" (khi). King T'ang and the Shang dynasty were responsible for unifying China to form their first civilization.

Ancient Mesopotamia

Many scholars have concluded that the founders of the first Mesopotamian civilization were Black Sumerians. Mesopotamia was the Biblical land of Shinar (Sumer), which sprung up around 3000 B.C.

After deciphering the cuneiform script and researching ancient Mesopotamia for many years Henry Rawlinson (1810-1895) discovered that the founders of the civilization were of Kushite (Cushite) origin. He made it clear that the Semitic speakers of Akkad and the non-Semitic speakers of Sumer were both Black people who called themselves sag-gig-ga or "Black Heads."

John Baldwin wrote in his book *"PreHistoric Nations"* (1869): "The early colonists of Babylonia were of the same race as the inhabitants of the Upper Nile."

This was corroborated by other scholars including, Chandra Chakaberty, who asserted in his book *"A Study in Hindu Social Polity"* that "based on the statuaries and steles of Babylonia, the Sumerians were "of dark complexion (chocolate colour), short stature, but of sturdy frame, oval face, stout nose, straight hair, full head; they typically resembled the Dravidians, not only in cranium, but almost in all the details."

Sources:

(PreHistoric Nations by John D. Baldwin, New York: Harper & Brothers, 1869, pg. 192)
(A Study in Hindu Social Polity by Chandra Chakaberty, Delhi: Mittal

Publications, 1987, pg. 33)
(From Babylon to Timbuktu by Rudolph R. Windsor. Atlanta: Windsor's Golden Series, 2203)

****

Clyde A. Winters and Blacks in China

Taiwan's aboriginal clan

The skeletal remains from Southern China are predominately Negroid. The people of that era practiced single burials which is an African ritual. In northern China Blacks founded many civilizations. The three major empires of China were the Xia Dynasty (c.2205-1766 BC), Shang Yin Dynasty (c.1700-1050 BC) and the Zhou Dynasty. The Zhou dynasty was the first dynasty founded by the Mongoloid people in China called Hua (Who-aa). The founders of Xia and Shang came from the Fertile African Crescent by way of Iran. Chinese civilization began along the Yellow River. By 3500 BC. Blacks in China were raising silkworms and making silk. The culture hero Huang Di is a direct link of Africa. His name was pronounced in old Chinese Yuhai Huandi or "Hu Nak Kunte." He arrived in China from the west in 2282 BC and settled along the banks of the Loh River in Shanxi. This transliteration of Huandgi, to Hu Nak Kunte is interesting because Kunte is a common clan name among the Manding speakers. The Africans or Blacks that founded civilization in China were often called Li Min "black headed people" by the Zhou dynasts. This term has affinity to the Sumero-Akkadian term Sag- Gig-Ga "black headed people."

China was occupied predominately by Blacks from West Asia to China. Blacks were forced from East and Southeast Asia by the expansion of the Thai, Annamite, Bak and Hua Mongoloids. Blacks ruled China until around 1000-700 BC. Blacks of China were known in historical literature by many names, including Negro, Austroloid, Oceanean, etc. by the Europeans. The East Indians and Mongoloid groups had other names like Dara, Yneh-chih, Yaksha, Suka, K'un-lun, Lushana and Seythians.

Hawaii's last Queen and her Black brother
  (The Ancient Sea People)

1838 - 1913
Queen
Lili'uokalani

1891
Queen
Lili'uokalani
Hawaii's last
Queen

1874
Lili'uokalani's
brother
Kalakaua
becomes King

The original Black population lived in China and were the <u>Negritos</u> and Austroloid groups. After 5000 BC, Africoid people from Kush in Africa began to enter China and Central Asia from Iran, while another groups reached China by sea. This two-route migration of Blacks to China led to the development of southern and northern Chinese branches of Africoids. The Northern Chino-Africans were called Kui-shuang (Kushana) or Yueh-chih, while the southern tribes were called Yi and li-man Yueh and Man. In addition to the Yueh Tribes along the north east coastal region, Blacks also lived in Turkestand, Mongolia, Transoxiana, the Ili region and Xinjiang Province.

Fifty thousand (50,000) years ago, the earliest forms of man were believed to have migrated from the Asian Subcontinent to the Philippine Islands via land bridges formed during the Ice Age. People of the <u>Negrito Race</u> came to the Philippines. <u>Negritos</u> can be described as a generally under five feet tall, flat nosed, dark-skinned with curly brown hair.

A number of factors lend support to the out-of-Africa hypothesis. The <u>Negrito</u> *look* African. Their skin color is light by African standards (though pygmy skin color is also lighter than their Bantu neighbors), but the rest of their physiology appears African. An interesting detail is the fact that the <u>Negrito</u> *sit* like pygmies, with their legs stretched out straight in front of them; I know of no other people who sit that way. The socio-economic relationship between the <u>Negrito</u> and their neighbors is strikingly analogous to that found in Africa. The relic populations of Vedda peoples found in Indonesia, Sri

Lanka, and Arabia. The logical explanation for the presence of Blacks worldwide is part of a great migration.

In Southeast Asia and southern China, ancient skeletal remains represented the earliest inhabitants to be Austroloids and Negrillo/Negrito. By the beginning of the Present (Holocene) Period the population in China could be differentiate, and placed into categories designating Mongoloid in the north, and Oceanic on Black Races in the south. Below is an excerpt from Nsaka Sesepkekiu - Student of African and Asian Studies - University of the West Indies - Trinidad and Tobago who validates what is being said:

The original, first, native, primitive inhabitants of China were black Africans who arrived there about 100,000 years ago and dominated the region until a few thousand years ago when the Mongol advance into that region began. These Africans who fled the Mongol onslaught can still be found in South East Asia and the Pacific Islands misnomer Nigritos or "small black men." The Agta of the Philippines is one such example. Indeed archeology, forensic and otherwise confirm that China's first two dynasties, the Xia and the Ch'ang/Sh'ang, were largely Black African with an Australoid, called "Madras Indian" or "Chamar" in Trinidad, present in small per-centages. These Africans would carry an art of fighting developed in the Horn of Africa into China which today we call martial arts: Tai Chi, Kung fu and Tae Kwon Do. Even the oracle of the I-Ching came with a later African group, the Akkadians of Babylon.

Around 500 BCE an African living in India called Gautama would establish a religion called Buddhism which would come to dominate Chinese thought. Any one who is in doubt should consult Geoffrey Higgins's Anacalypsis, Albert Churchward's _Origin and Development of Religions_, Gerald Massey's, _Egypt the Light of the World_, Runoko Rashidi's _African Presence in Early Asia_ and J A Roger's _Sex and Race Vol. 1_. Many Africans survived the Mongol invasion into the twentieth century only to be exterminated by Chairman Mao's program of Cultural cleansing. Under this program millions of Africans and Afro-Asians were killed from 1951-1956. Contribute we still did, giving the People's Republic of China its first Chief Minister in the name of Eugene Chen, a Trinidadian of George Street, Port-of-Spain, who was of an African mother and a Chinese father.

The facts are well recorded in African, East Indian and African-American history books. China also has a series of pyramids and groups of people "minorities" in the South such as the Moi of Vietnam and the Nakhis of Southern China. Cheikh Diop's points are well made when he stressed that the Yellow Race has racial characteristics of

both Negroid and Caucasian Races. The mixture of the two races created the Yellow Race. Below are pictures of Black Chinese from:
http://community2.webtv.net/BARNUBIANEMPIRE/BLACKPEOPLEBLACK/index.html

 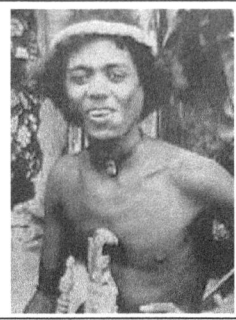 

This picture below of Buddha portrays him as a Negroid individual with kinky, coiled hair, flat nose, full lips. According to sources from India (M. Gopinath, "Nagaloka: The Fractured History and Forgotten Glory of the Bahujan Indians," published by Dalit Sahitya Sanghatane, No. 8, North Street, Neelasandra, Bangalore, India - 560 047 India) Buddha was, "an Enlightened Master from the Sakya clan of Naga Race, he was the first man on earth to preach the great principles off equality, liberty and fraternity. He made the Nagas to realize their own "mind power" as against the "mantra power".

Indo-Europeans

Several years ago a major television network covering then President Clinton's trip to sub-Saharan Africa captured him being greeted by dignitaries. A young girl dressed in a traditional skirt made from reeds was doing a dance with movements reminiscent of the Hawaiian Hula Dance.

The dance focused on shaking of the belly. This dance was once common in southern Africa countries and once widespread on the continent before the influences of fundamentalist Islam and Christianity. The dancing was once common among Nilotic people. In Egypt the category of traditional dances is called balladi, meaning simply "dances from the land". The dance has been preserved in Africa from the days of the Pharaohs. Could this same belly dance be *reminiscent of Northern Africa and Hawaii--since the traveling of Africans "FROM AFRICA--NOT TO AFRICA!"*

Now with that said let's tackle the once scarce history of the Indo-Europeans and their ploy to capture and claim what is Black African. Indo-European means the following:

1. A family of languages consisting of most European languages as well as those of Iran, the Indian subcontinent, and other parts of Asia. Proto-Indo-European also called Indo-Germanic.

2. Aryan--applied to the languages of India and Europe which are derived from the prehistoric Aryan language; also, pertaining to the people or nations who speak these languages; as, the Indo-European or Aryan family.

3. A member of any of the peoples speaking an Indo-European language. A member of one of the Caucasian Races of Europe or India speaking an Indo-European language.

And I quote from the Encyclopedia: "There is little known of the Indo-European homeland, but what is known comes from the words that can be reconstructed from their variants in the Indo-European Languages." As with the continental Old European civilizations, the Indo-European tribes started arriving in the Middle East--remember, once called "Africa" only very shortly after the first Old European society had been established in that region, in the so-called fertile river valley between **the Tigris and Euphrates Rivers in present day Iraq.**

Blacks are the first people to inhabit Arabia (the Adites and Cushites), Iraq (Cushites, Elamites), Iran (Elamites, Susians, Cushites). The first whites to enter the Black regions of Mesopotamia were the Gutis who swept from the Zagros Mountains and attempted to Destroy Black Mesopotamia. After many centuries, these invaders and some of

the Blacks mixed to create the Assyrians, Babylonians and the present population of the region. Yet, Black features and faces are still indigenous in parts of Iran and Iraq and South Arabia

Samoa

Solomon Islands

North Africa - 1890

North Africa - 1894

Samoa

Samoa

Moors' Influences and Contributions to Europe

by Ivan Van Sertima

A distinction should be drawn between the Classical Renaissance of Europe, which mainly relates to its literature and art, and the scientific Renaissance, which began to bud and flower in the 12th and 13th centuries. Jose Pimienta-Bey deals primarily with the Moorish stimulus for the latter. He sets out to prove in his essay, and does so with a formidable body of evidence, that the foundation of much of medieval western science and its academies was built up upon the transmissions, refinements and discoveries of the Arabs and the Moors.

Moorish influence came primarily to the West by way of the Iberian peninsula (renamed al-Andalus by the Moors). Bey provides us with a detailed examination of Western Europe's scholarly relations with Spain. Translation, of course, played the major role in this diffusion of the sciences. The schools of translation were like the bridges between the Muslim and Christian scholars. Chief among these was the school of translators founded at Toledo by Alfonso X during the thirteenth century. Translations from Arabic (the medieval language of science) into Latin, the classical European language, had been going on since the tenth century. Centers of translation sprang up all over Christian Europe - Barcelona, Tarazona, Iron, Segovia, Pamplona, Toulouse, Beziers, Narbonne, Marseilles. Bologna, Salerno and Paris made extensive use of Moorish scientific treatises. The translations from the Arabic provided links between Spain, Portugal, France, Italy and England. Alphonso X promoted Moorish erudition at every opportunity. The first university of Christian Spain was founded at Valencia by Alfonso VIII in the 13th century and the teachers employed were the Muslims and the Jews. Nearly all the major universities in Europe sprung up around the same time, beginning in the second half of the 12th century right up through the 13th, a span of about one hundred and fifty years, a period which coincides with the flowering of Moorish science and the establishment of centers in Europe to translate Moorish treatises from Arabic into Latin. In Italy we have Bologna, Padua, Naples, Rome; in France, Montpelier and Toulouse; in Portugal, Lisbon and Coimbra; in England, Oxford.

Several of the Moorish works in mathematics, astronomy and medicine became standard texts at these universities. For example, Judwal, a Moorish work in astronomy, became a standard text at Oxford. Frederick II founded a university at Naples in 1224 and there he established a curriculum which emphasized Moorish scholarship. Under him all theological studies ceased at
  Italian universities and Moorish medicine and law became the major disciplines.
  A curious schizophrenia developed among the Catholics in relation to Moorish science and knowledge. On the one hand they were very much aware of the superior knowledge of the Moors and they made efforts to acquire that knowledge so that they would not be left too far behind. At the same time they strove desperately to keep it away from the common people and even, at times, to vilify it so that it would not become a challenge to Catholicism. They were afraid that the Enlightenment, the new ideas that this new knowledge would bring, could affect the populace. So that, even though they were given the keys to the inner sanctum, they kept the cage closed to the masses. Into Europe came the advances of an empire more immense than those of either Alexander the Great or Rome at its height. Rice was introduced into Europe by the Moors in the tenth century, cotton by the ninth. A Moorish botanist, Ibn Bassal, partitioned the land into ten different classes, according to particular characteristics, and taught the farmers ways of increasing the fertility of their plots. Surveys were done to locate sweet water below the earth. Widespread use was made of the waterwheel which the Moors had introduced into Spain. The Romans also knew of this but they had used it very little. The Moors also dug canals and channels to irrigate the farmlands and provide water for the thousands of houses and mosques and palaces and public baths. They not only increased the fertility of the soil with their new methods and tools and plants and manures but they also ushered in the sciences of food preservation and storage. They could store wheat for as long as one hundred years. Their methods of drying enabled such food as figs, plums, cherries and apples to remain edible for several years. They have left the voiceprint of their language on the things they introduced. A lot of Arabic words have entered general usage as a result of the Moorish invasion of Europe. Bey cites coffee, sugar, rice, cotton, lemon, syrup, soda, alcohol, alkali, cipher, algebra, arsenal, admiral, alcove, magazine.
  Let me add a few to this list, selected from my own work on pre-Columbian navigation and the transfer of plants: anchor (from angar) caravel (from caravos) tobacco (from dubbaq and a series of taba and

tabgha words).1 I Also, the technical terms for the astrolabe (an astronomical device invented by the Moors) still retain their Arabic names.

But technology in itself is not the only arbiter of civilization. It is important to note a benign African influence on the way Islam operated in Spain, particularly in relationship to women. Ibn Battuta, the Arab traveller and writer, first commented with astonishment on the level of freedom and equality of Muslim women in the African town of Walata.12 It was the same in Moorish Spain. Unique among Islamic nations, women enjoyed more societal freedoms than in any part of the Islamic world. They moved freely in public and engaged in various gatherings. The practice of purdah was almost entirely ignored in Moorish Spain. Even a daughter of a 12th century Caliph had a total disregard for the veil.

A question that has always haunted me is the reason for Europe's Dark Age. Why did Europe fall into such darkness after all it had received from the Greeks who had taken so much from, and added what they could to, the Egyptian sciences? G.G.M. James, in The Stolen lrgacy, answers this question. James had pointed out that the edicts of Theodosius in the 4th century closed down the temples of the Egyptian mysteries as well as the philosophical schools of Greece. The emperor Justinian in 529 A.D. followed in the same path of Theodosius. Thus an intellectual darkness descended over Christian Europe and the entire Greco-Roman world. It lasted for centuries. But I feel James exaggerates when he claims that "the Greeks showed no creative powers and were unable to improve on the knowledge they received."13 His point about their borrowings is well made but this kind of chauvinistic remark is quite unnecessary. There is no need to suggest the Greeks were dumb and could make no improvements whatever on what they learnt. If that were true, the influence of the Egyptians would have been negligible.

But James makes an even more important point which I have not seen repeated elsewhere. It is the missing link in the drama of Moorish scientific ascendancy in the Middle Ages. Eurocentric historians had argued that the Arabs were merely transmitting the Greek heritage lost to Europe during its Dark Age. Even Arabs were made to believe that and to assume that they were standing on the shoulders of Greek giants. By the time they attacked Egypt, Europeans had long been in charge of that defeated country. The Arabs seemed to forget that their conquest of Egypt had been made easy by the resentment of the Egyptians against their Byzantine overlords. We know far more today about the enonnous debt Greek science owes to Egypt (see my essay

"The Egyptian Precursor" in this issue). But what was little suspected was that Greece was not the only conduit of Egyptian scientific genius to the Arab world. James provides evidence that there were Egyptians fleeing their country in large numbers during the Persian, Greek and Roman invasions, fleeing not only to the desert and mountain regions but also to adjacent lands in Africa-Arabia and Asia Minor, where, "they lived and secretly developed

the teachings which belonged to their Mystery System. In the eighth century A.D. the Moors of North Africa invaded Spain and took with them the Egyptian culture which they had preserved.

Knowledge in the ancient days was centralized, that is, it belonged to a common parent and system – the Wisdom System or Mysteries of Egypt, which the Greek used to call Sophia." Whatever we may say of these great scientific advances, there is something that we cannot gloss over and which unfortunately we must mention in our uncompromising quest for the truth of history. Some despots and merchants did trade in slaves during part of the Moorish occupation of al-Andalus. Most of these, before the European slave trade were European slaves. It has been said that slavery among the Muslims and slavery among the Catholics had important differences. Bey quotes Joseph O'Callahan who, in The History of Medieval Spain, makes it clear that "owners did not possess the power of life and death over them nor could they inflict excessive punishment. Slaves had rights and they could actually seek assistance if they were exceedingly maltreated." On this matter Bey comments "any student of American history knows that this was far from the case regarding the British and United States system of enslavement. The enslaved African was a non-human legally designated as 'property'."

Slavery, regardless of these qualifications, can never be condoned or forgiven. But it was not central to their system: it was marginal. I think it should also be pointed out, contrary to myths about the Muslims, that they did not force their religion down the throats of the Christians. John Jackson, in an informative chapter on the Moors, in his book Introduction to African Civilizatinn, 14 shows us how Christian, Jew and Muslim were treated with equal respect during the dynasty of the Ummayads. We have been given no evidence that this changed dramatically in later Muslim dynasties.

The slave trade in this time was not a state institution. It was like the lucrative drug enterprise of today - a large but lawless thing, sometimes indulged in by bad rulers but not a keystone of the system, as it was later to become in the Euro- Christian world. The Moors, let it be said, did not suppress the languages of the people of Al-Andalus, *they did not outlaw their sacred customs, they did not turn Iberia into a sweat-shop, its fertile lands a mere source of raw materials for the Muslim international elite. They did not destroy their legal system, rob them of their political rights, deny them their claim to humanity. The one thing they did insist on, was a say in the election of the Catholic bishops since the rival power of the church could undermine Muslim power and authority*

The world changed dramatically in 1492, not only because Columbus stumbled in the direction of the Americas, using the magnet of a myth to draw millions behind him, but because that was the very year the Moors were defeated. It is not an accident that it is Spain and Portugal who spearheaded the movement in this direction. It was on Ian.2, 1.492 that the African leader, Abu Abdi-Llah, otherwise known as Boabdil, surrendered to the Spanish. Jan Carew compares the illiteracy of the Christian Europeans to the learning and erudition of the Moors of that time. The comparison is so startling, his comment is worth quoting.

"At a time when the most insignificant provinces of Moorish Spain contained libraries running into thousands of volumes, the cathedrals, monasteries and palaces of Leon, under Christian rule, numbered books only by the dozen. The paltry number of texts the Christians did possess were almost all devotional or liturgical."

The narrowness of vision this produced among leaders of the church and state was to have catastrophic effects. It led to the massive burning of African and Arab books under the order of Cardinal Ximenes de Cisneros. It inspired a similar bonfire of the books of native Americans. Bishop de Landa exhorted his followers in the Yucatan "Burn them all - they are works of the devil." The destruction of the Moorish libraries was particularly vicious because it was not only inspired by religious narrowness and bigotry. Hatred of the dark invaders kindled the bonfires. The Church at that time too saw most of this foreign learning as something evil, even demonic. The number system that we use today, for example, brought in by the Moors from India, was seen as late as the 17th century in some parts of Europe as signs of the devil. It became a religious mission for men like Ximenes and his successors to erase from history all memory of the Moors. Ximenes even induced the Spanish sovereigns to outlaw the public baths, making cleanliness antithetical to godliness.

Fortunately for the scientific renaissance, key Moorish works had already been translated and circulated, even smuggled secretly into the academies, significant seminal inventions introduced and established before these barbaric attempts at an intellectual holocaust.

****

Ancient African Kings of India, by Dr. Clyde Winters

Ethiopians have had very intimate relations with Indians. In fact, in antiquity the Ethiopians ruled much of India. These Ethiopians were called the Naga. It was the Naga who created Sanskrit.

A reading of ancient Dravidian literature which dates back to 500 BC, gives us considerable information on the Naga. In Indian tradition the Naga won central India from the Villavar (bowmen) and Minavar (fishermen).

The Naga were great seamen who ruled much of India, Sri Lanka and Burma. To the Aryans they described as half man and snake. The Tamil knew them as warlike people who used the bow and noose.

The earliest mention of the Naga, appear in the Ramayana, they are also mentioned in the Mahabharata. In the Mahabharata we discover that the Naga had the capital city in the Dekkan, and other cities spread between the Jumna and Ganges as early as 1300 BC. The Dravidian classic, the Chilappathikaran made it clear that the first great kingdom of India was Naganadu.

The Naga probably came from Kush-Punt/Ethiopia. The Puntites were the greatest sailors of the ancient world. In the Egyptian inscriptions there is mention of the Puntite ports of Outculit, Hamesu and Tekaru, which corresponds to Adulis, Hamasen and Tigre.

In Sumerian text, it is claimed that the Puntites traded with the people of the Indus Valley or Dilmun. According to S.N. Kramer in The Sumerians, part of Punt was probably called Meluhha, and Dilmun was probably the ancient name of the Indus Valley. (Today some scholars maintain that Oman, where we find no ancient cities was Dilmun and the Indus Valley may have been Meluhha).

Ancient Ethiopian traditions support the rule of Puntites or Ethiopians of India. In the Kebra Nagast, we find mention of the Arwe kings who ruled India. The founder of the dynasty was Za Besi Angabo. This dynasty according to the Kebra Nagast began around 1370 BC. These rulers of India and Ethiopia were called Nagas. The Kebra Nagast claims that "Queen Makeda had servants and merchants; they traded for her at sea and on land in the Indies and Aswan". It also says that her son Ebna Hakim or Menelik I, "made a campaign in the Indian Sea; the king of India made gifts and donations and prostrated himself before him". It is also said that "Menalik ruled an empire that extended from the rivers of Egypt (Blue Nile) to the west and from the south Shoa to eastern India", according to the Kebra Nagast. The Kebra Nagast identification of an eastern Indian empre ruled by the Naga, corresponds to the Naga colonies in the Dekkan, and on the East coast between the Kaviri and Vaigai rivers.

"…The darkest man is here the most highly esteemed and considered better than the others who are not so dark. Let me add that in very truth these people portray and depict their gods and their idols black and their devils white as snow. For they say that God and all the saints are black and the devils are all white. That is why they portray them as I have described." – Marco Polo, after visiting the Pandyan Kingdom in 1288

Malik Andil Khan Sultan (reigned from 893 AD -895 AD)

Little is known about Shahzada Khoja Barbak – the Ethiopian that conquered the Bengal Kingdom and established the Habshi dynasty in 1487. But we know that he was a Siddi.

Jamal al-Din Yaqut (ca 1200)

Jamal began his rise to power in Delhi as a habshi, one of many enslaved Africans of East African descent frequently employed by Muslim monarchs as mercenaries and members of royal security teams. Shortly after his employ began, the then reigning sovereign Queen Raziya (1236- 1240) the first female monarch of Delhi took a liking to him. He was subsequently promoted to a royal courtier and later rose to occupy the important post of superintendent of the royal stables.

Malik Sarwar (1304 – 1403)

Malik Sarwar, also described as a Habashi, became the governor of Jaunpur, a sultanate close to Delhi. Under the title of Malik-us-Shark (king of the east) he captured the Jaunpur province. According to the History of Medieval India, Part I (S.Chand& Co, 2007), "In 1389, Malik Sarwar received the title of Khajah-i-Jahan. In 1394, he was appointed as the governor of Jaunpur and received his title of Malik-us-Sharq from Sultan Nasiruddin Mahmud Shah II Tughluq (1394 – 1413)

Malik Ambar (1550 - ?)

One of the most famous among the Indo-Africans was the celebrated Malik Ambar (1550-1626). Malik Ambar, whose original name was Shambu was born around 1550 in Harar, Ethiopia. After his arrival in India, he was able to raise a formidable army and achieve great power in the west Indian realm of Ahmadnagar.

Ambar was a brilliant diplomat, tactician, and administrator. In 1590, Ambar broke away from Bijapur and built an independent mercenary army of over 1500 African, Arab and local Dakani men.

He eventually joined the state of Ahmadnagar and later imprisoned King Murtaza II, naming himself regent minister.

Ambar also organized a 60,000 horse army and successfully beat back the Moguls for the next 20 years. The Moguls could not conquer Dakan until after his death.

Black Contributions to the Americas

BLACK CIVILIZATIONS OF ANCIENT AMERICA (MUULAN), MEXICO

Gigantic stone head of Negritic African during the Olmec (Xi) Civilization

By Paul Barton

The earliest people in the Americas were people of the Negritic African race, who entered the Americas perhaps as early as 100,000

years ago, by way of the bering straight and about thirty thousand years ago in a worldwide maritime undertaking that included journeys from the then wet and lake filled Sahara towards the Indian Ocean and the Pacific, and from West Africa across the Atlantic Ocean towards the Americas.

According to the Gladwin Thesis, this ancient journey occurred, particularly about 75,000 years ago and included Black Pygmies, Black Negritic peoples and Black Australoids similar to the Aboriginal Black people of Australia and parts of Asia, including India.

Ancient African terracotta portraits 1000 B.C. to 500 B.C.

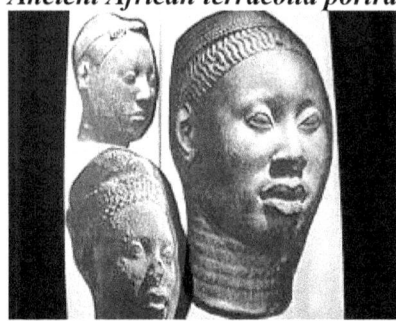

Recent discoveries in the field of linguistics and other methods have shown without a doubt, that the ancient Olmecs of Mexico, known as the Xi People, came originally from West Africa and were of the Mende African ethnic stock. According to Clyde A. Winters and other writers (see Clyde A. Winters website), the Mende script was discovered on some of the ancient Olmec monuments of Mexico and were found to be identical to the very same script used by the Mende people of West Africa. Although the carbon fourteen testing date for the presence of the Black Olmecs or Xi People is about 1500 B.C., journies to the Mexico and the Southern United States may have come from West Africa much earlier, particularly around five thousand years before Christ. That conclusion is based on the finding of an African native cotton that was discovered in North America. It's only possible manner of arriving where it was found had to have been through human hands. At that period in West African history and even before, civilization was in full bloom in the Western Sahara in what is today Mauritania. One of Africa's earliest civilizations, the Zingh Empire, existed and may have lived in what was a lake filled, wet and fertile Sahara, where ships criss-crossed from place to place.

ANCIENT AFRICAN KINGDOMS PRODUCED
OLMEC TYPE CULTURES

The ancient kingdoms of West Africa which occupied the Coastal forest belt from Cameroon to Guinea had trading relationships with other Africans dating back to prehistoric times. However, by 1500 B.C., these ancient kingdoms not only traded along the Ivory Coast, but with the Phoenicians and other peoples. They expanded their trade to the Americas, where the evidence for an ancient African presence is overwhelming. The kingdoms which came to be known by Arabs and Europeans during the Middle Ages were already well established when much of Western Europe was still inhabited by Celtic tribes. By the 5th Century B.C., the Phoenicians were running comercial ships to several West African kingdoms. During that period, iron had been in use for about one thousand years and terracotta art was being produced at a great level of craftsmanship. Stone was also being carved with naturalistic perfection and later, bronze was being used to make various tools and instruments, as well as beautifully naturalistic works of art.

The ancient West African coastal and interior Kingdoms occupied an area that is now covered with dense vegetation but may have been cleared about three to four thousand years ago. This includes the regions from the coasts of West Africa to the South, all the way inland to the Sahara. A number of large kingdoms and empires existed in that area. According to Blisshords Communications, one of the oldest empires and civilizions on earth existed just north of the coastal regions into what is today Mauritania. It was called the Zingh Empire and was highly advanced. In fact, they were the first to use the red, black and green African flag and to plant it throughout their territory all over Africa and the world.

The Zingh Empire existed about fifteen thousand years ago. The only other civilizations that may have been in existance at that period in history were the Ta-Seti civilization of what became Nubia-Kush and the mythical Atlantis civilization which may have existed out in the Atlantic, off the coast of West Africa about ten to fifteen thousand years ago. That leaves the question as to whether there was a relationship between the prehistoric Zingh Empire of West Africa and the civilization of Atlantis, whether the Zingh Empire was actually Atlantis, or whether Atlantis if it existed was part of the Zingh empire. Was Atlantis, the highly technologically sophisticated civilization an extension of Black civilization in the Meso-America and other parts of the Americas?

From the archeological evidence gathered both in West Africa and Meso-America, there is reason to believe that the African Negritics who founded or influenced the Olmec civilization came from West Africa. Not only do the collosol Olmec stone heads resemble Black Africans from the Ghana area, but the ancient religious practices of the Olmec priests was similar to that of the West Africans, which included shamanism, the study of the Venus complex which was part of the traditions of the Olmecs as well as the Ono and Dogon People of West Africa. The language connection is of significant importance, since it has been found out through decipherment of the Olmec script, that the ancient Olmecs spoke the Mende language and wrote in the Mend script, which is still used in parts of West Africa and the Sahara to this day.

ANCIENT TRADE BETWEEN THE AMERICAS AND AFRICA

The earliest trade and commercial activities between prehistoric and ancient Africa and the Americas may have occurred from West Africa and may have included shipping and travel across the Atlantic. The history of West Africa has never been properly researched. Yet, there is ample evidence to show that West Africa of 1500 B.C. was at a level of civilization approaching that of ancient Egypt and Nubia-Kush. In fact, there were similarities between the cultures of Nubia and West Africa, even to the very similarities between the smaller scaled hard brick clay burial pyramids built for West African Kings at Kukia in pre-Christian Ghana and their counterparts in Nubia, Egypt and Meso-America.

Although West Africa is not commonly known for having a culture of pyramid-building, such a culture existed although pyramids were created for the burial of kings and were made of hardened brick. This style of pyramid building was closer to what was built by the Olmecs in Mexico when the first Olmec pyramids were built. In fact, they were not built of stone, but of hardened clay and compact earth.

Still, even though we don't see pyramids of stone rising above the ground in West Africa, similar to those of Egypt, Nubia or Mexico, or massive abilisks, collosal monuments and structures of Nubian and Khemitic or Meso-American civilization. The fact remains, they did exist in West Africa on a smaller scale and were transported to the Americas, where conditions such as an environment more hospitable to *building and free of detriments such as malaria and the tsetse fly, made it much easier to build on a grander scale.*

Meso-American pyramid with stepped appearance, built about 2500 years ago

Stepped Pyramid of Sakkara, Egypt, built over four thousand years ago, compare to Meso-American pyramid

Large scale building projects such as monuent and pyramid building was most likely carried to the Americas by the same West Africans who developed the Olmec or Xi civilization in Mexico. Such activities would have occurred particularly if there was not much of a hinderance and obstacle to massive, monumental building and construction as there was in the forest and malaria zones of West Africa. Yet, when the region of ancient Ghana and Mauritania is closely examined, evidence of large prehistoric towns such as Kukia and others as well as various monuments to a great civilization existed and continue to exist at a smaller level than Egypt and Nubia, but significant enough to show a direct connection with Mexico's Olmec civilization.

The similarities between Olmec and West African civilization includes racial, religious and pyramid bilding similarities, as well as the similarities in their alphabets and scripts as well as both cultures speaking the identical Mende language, which was once widespread in the Sahara and was spread as far East as Dravidian India in prehistoric times as well as the South Pacific.

During the early years of West African trade with the Americas, commercial seafarers made frequent voyages across the Atlantic. In fact, the oral history of a tradition of seafaring between the Americas and Africa is part of the history of the Washitaw People, an aboriginal Black nation who were the original inhabitants of the Mississippi Valley region, the former Louisiana Territories and parts of the Southern United States. According to their oral traditions, their ancient ships criss-crossed the Atlantic Ocean between Africa and the Americas on missions of trade and commerce.

Some of the ships used during the ancient times, perhaps earlier than 7000 B.C. (which is the date given for cave paintings of the drawings and paintings of boats in the now dried up Sahara desert) are similar to ships used in parts of Africa today. These ships were either made of papyrus or planks lashed with rope, or hollowed out tree trunks.

These ancient vessels were loaded with all type of trade goods and not only did they criss-cross the Atlantic but they traded out in the Pacific and settled there as well all the way to California.

In fact, the tradition of Black seafarers crossing the Pacific back and forth to California is much older than the actual divulgance of that fact to the first Spanish explorers who were told by the American Indians that Black men with curly hair made trips from California's shores to the Pacific on missions of trade.

On the other hand, West African trade with the Americas before Columbus and way back to proto historic times (30,000 B.C. to 10,000 B.C.), is one of the most important chapters in ancient African history. Yet, this era which begun about 30,000 years ago and perhaps earlier (see the Gladwin Thesis, by C.S.Gladwin, Mc Graw Hill Books), has not been part of the History of Blacks in the Americas. Later on in history, particularly during the early Bronze Age.

However, during the latter part of the Bronze Age, particularly between 1500 B.C. to 1000 B.C., when the Olmec civilization began to bloom and flourish, new conditions in the Mediterranean made it more difficult for West Africans to trade by sea with the region, although their land trade accross the Sahara was flourishing. By then, Greeks, Phoenicians, Assyrians, Babylonians and others were trying to gain *control of the sea routes and the trading ports of the region. Conflicts in the region may have pushed the West Africans to strengthen their trans-Atlantic trade with the Americas and to explore and settle there.*

Ancient sea-going vessel used by the Egyptians and Nubians in ancient times.

West African Trade and Settlement in the Americas Increases Due to Conflicts in the Mediterranean

The flowering of the Olmec Civilization occurred between 1500 B.C. to 1000 B.C., when over twenty-two collosal heads of basalt were carved representing the West African Negritic racial type.

This flowering continued with the appearance of "Magicians," or Shamanistic Africans who observed and charted the Venus planetary complex (see the pre-Christian era statuette of a West African Shaman in the photograph above) These "Magicians," are said to have entered Mexico from West Africa between 800 B.C. to 600 B.C. and were speakers of the Mende language as well as writers of the Mende script or the Bambara script, both which are still used in parts of West Africa and the Sahara.

These Shamans who became the priestly class at Monte Alban during the 800's to 600's B.C. (ref. The History of the African-Olmecs and Black Civilization of the Americas From Prehistoric Times to the Present Era), had to have journied across the Atlantic from West Africa, for it is only in West Africa, that the religious practices and astronomical and religious practices and complex (Venus, the Dogon Sirius observation and the Venus worship of the Afro-Olmecs, the use of the ax in the worship of Shango among he Yoruba of West Africa and the use of the ax in Afro-Olmec worship as well as the prominence of the thunder God later known as Tlalock among the Aztecs) are the same as those practiced by the Afro-Olmec Shamans. According to Clyde Ahmed Winters (see "Clyde A. Winters" webpage on "search."

Thus, it has been proven through linguistic studies, religious similarities, racial similarities between the Afro-Olmecs and West Africans, as well as the use of the same language and writing script, that the Afro-Olmecs came from the Mende-Speaking region of West Africa, which once included the Sahara.

Sailing and shipbuilding in the Sahara is over twenty thousand years old. In fact, cave and wall paintings of ancient ships were displayed in National Geographic Magazine some years ago. Such ships which carried sails and masts, were among the vessels that swept across the water filled Sahara in prehistoric times. It is from that shipbuilding tradition that the Bambara used their knowledge to build Thor Hayerdhal's papyrus boat Ra I which made it to the West Indies from Safi in Morroco years ago. The Bambara are also one of the West African nationalities who had and still have a religious and astronomical complex similar to that of the ancient Olmecs, particularly in the area of star gazing.

A journey across the Atlantic to the Americas on a good current during clement weather would have been an easier task to West Africans of the Coastal and riverine regions than it would have been through the use of caravans criss-crossing the hot by day and extremely cold by night Sahara desert. It would have been much easier to take a well made ship, similar to the one shown above and let the currents take it to the West Indies, and may have taken as long as sending goods back and forth from northern and north-eastern Africa to the interior and coasts of West Africa's ancient kingdoms. Add to that the fact that crossing the Sahara would have been no easy task when obsticales such as the hot and dusty environment, the thousands of miles of dust, sand and high winds existed. The long trek through the southern regions of West Africa through vallies, mountains and down the many rivers to the coast using beasts of burden would have been problematic particularly since malaria mosquitoes harmful to both humans and animals would have made the use of animals to carry loads unreliable.

Journeys by ship along the coast of West Africa toward the North, through the Pillars of Heracles, eastward on the Mediterran to Ports such as Byblos in Lebanon, Tyre or Sydon would have been two to three times as lengthy as taking a ship from Cape Verde, sailing it across the Atlantic and landing in North-Eastern Brazil fifteen hundred miles away, or Meso America about 2400 miles away. The distance in itself is not what makes the trip easy. It is the fact that currents which *are similar to gigantic rivers in the ocean, carry ships and other vessels from West Africa to the Americas with relative ease.*

West Africans during the period of 1500 B.C. to 600 B.C. up to 1492 A.D. may have looked to the Americas as a source of trade, commerce and a place to settle and build new civlilzations. During the period of 1500 B.C. to 600 B.C., there were many conflicts in the Mediterranean involving the Kushites, Egyptians, Assyrians, Phoenicians, Sea Peoples, Persians, Jews and others. Any kingdom or nation of that era who wanted to conduct smoothe trade without complications would have tried to find alternative trading partners. In fact, that was the very reason why the Europeans decided to sail westwared in their wearch for India and China in 1492 A.D. They were harrassed by the Arabs in the East and had to pay heavy taxes to pass through the region.

Still, most of the Black empires and kingdoms such as Kush, Mauri, Numidia, Egypt, Ethiopia and others may have had little difficulty conducting trade among their neighbors since they also were among the major powers of the region who were dominant in the Mediterranean.

South of this northern region to the south-west, Mauritania (the site of the prehistoric Zingh Empire) Ghana, and many of the same nationalities who ushered in the West African renaissance of the early Middle Ages were engaged in civilizations and cultures similar to those of Nubia, Egypt and the Empires of the Afro-Olmec or Xi (Shi) People.

Nubian-Kushite King and Queen (circa 1000 B.C.)

It is believed that there was a Nubian presence in Mexico and that the West African civilizations were related to that of the Nubians, despite the distance between the two centers of Black civilization in Africa.
There is no doubt that in ancient times there were commercial ties between West Africa and Egypt. In fact, about 600 B.C., Nikau, a Pharaoh of Egypt sent ships to circumnavigate Africa and later on about 450 B.C., Phoenicians did the same, landing in West Africa in

the nation now called Cameroon. There they witnessed what may have been the celebration of a Kwanza-like harvest festival, where "cymbals, horns," and other instruments as well as smoke and fire from buring fields could be seen from their ships.

At that period in history, the West African cultures and civilizations, which were offshoots of much earlier southern Saharan cultures, were very old compared to civilizations such as Greece or Babylon. In fact, iron was being used by the ancient West Africans as early as 2600 years B.C. and was so common that there was no "bronze age" in West Africa, although bronze was used for ornaments and instruments or tools.

A combination of Nubians and West Africans engaged in mutual trade and commerce along the coasts of West Africa could have planned many trips to and from the Americas and could have conducted a crossing about 1500 B.C. and afterwards. Massive sculptures of the heads of typical Negritic Africans were carved in the region of South Mexico where the Olmec civilization flourished. Some of these massive heads of basalt contain the cornrow hairstyle common among West African Blacks, as well as the kinky coiled hair common among at least 70 percent of all Negritic people, (the other proportion being the Dravidian Black race of India and the Black Australoids of Australia and South Asia).

Collossol Afro-Olmec head of basalt wearing Nubian type war helmet, circa 1100 B.C.

Afro-Olmecs Came from the Mende Regions of West Africa

Although archeologists have used the name "Olmec," to refer to the Black builders of ancient Mexico's first civilizations, recent discoveries have proven that these Afro-Olmecs were West Africans of the Mende language and cultural group. Inscriptions found on ancient monuments in parts of Mexico show that the script used by the ancient Olmecs was identical to that used by the ancient and modern Mende-speaking peoples of West Africa. Racially, the collosal stone heads are identical in features to West Africans and the language deciphered on Olmec monuments is identical to the Mende language of West Africa, (see Clyde A. Winters) on the internet.

The term "Olmec" was first used by archeologists since the giant stone heads with the features of West African Negritic people were found in a part of Mexico with an abundance of rubber trees. The Maya word for rubber was "olli, and so the name "Olmec," was used to label the Africoid Negritic people represented in the faces of the stone heads and found on hundreds of terracotta figurines throughout the region.

Yet, due to the scientific work done by deciphers and linguists, it has been found out that the ancient Blacks of Mexico know as Olmecs, called themselves the Xi People (She People).

Apart from the giant stone heads of basalt, hundreds of terracotta figurines and heads of people of Negritic African racial reatures have also been found over the past hundred years in Mexico and other parts of Meso-America as well as the ancient Black-owned lands of the Southern U.S. (Washitaw Proper,(Texas, Louisiana, Mississippi, Oklahoma, Arkansas), South America's Saint Agustin Culture in the nation of Colombia, Costa Rica, and other areas) the "Louisiana Purchase," lands, the south-eastern kingdom of the Black Jamassee, and other places including Haiti, see the magazine Ancient American).

Various cultural clues and traces unique to Africa as well as the living descendants of prehistoric and ancient African migrants to the Americas continue to exist to this very day. The Washitaw Nation of Louisiana is one such group (see www.Hotep.org), the Garifuna or Black Caribs of the Caribbean and Central America is another, the descendants of the Jamasse who live in Georgia and the surrounding states is another group. There are also others such as the Black Californian of Queen Calafia fame (the Black Amazon Queen mentioned in the book Journey to Esplandian, by Ordonez de Montalvo during the mid 1500's).

Cultural artefacts which connect the ancient Blacks of the Americas with Africa are many. Some of these similarities can be seen in the stone and terracotta works of the ancient Blacks of the Americas. For example, the African hairline is clearly visible in some stone and terracotta works, including the use of cornrows, afro hair style, flat "mohawk" style similar to the type used in Africa, dreadlocks, braided hair and even plain kinky hair. The African hairline is clearly visible on a fine stone head from Veracruz Mexico, carved between 600 B.C. to 400 B.C., the Classic Period of Olmec civilization. That particular *statuette is about twelve inches tall and the distance from the head to the chin is about 17 centemeters. Another head of about 12 inches, not only posesses Negroid features, but the hair design is authentically West African and is on display at the National Museum of Mexico. This terracotta Africoid head also wears the common disk type ear plugs common in parts of Africa even today among tribes such as the Dinka and Shilluk.*

One of the most impressive pieces of evidence which show a direct link between the Black Olmec or Xi People of Mexico and West Africans is the presence of scarification marks on some Olmec terracotta sculpture. These scarification marks clearly indicate a West African Mandinka (Mende) presence in prehistoric and ancient Meso-America. Ritual scarification is still practiced in parts of Africa and among the Black peoples of the South Pacific, however the Olmec scarification marks are not of South Pacific or Melanesian Black origins, since the patterns used on ancient Olmec sculpture is still common in parts of Africa. This style of scarification tatooing is still used by the Nuba and other Sudanese African people. In fact, the face of a young girl with keloid scarification on here face is identical to the very same keloid tatoos on the face of an ancient Olmec terracotta head from ancient Mexico. Similar keloid tattoos also appear on the arms of some Sudanese and are identical to similar keloid scars on the arms of some clay figures from ancient Olmec terracotta figurines of Negroid peoples of ancient Mexico.

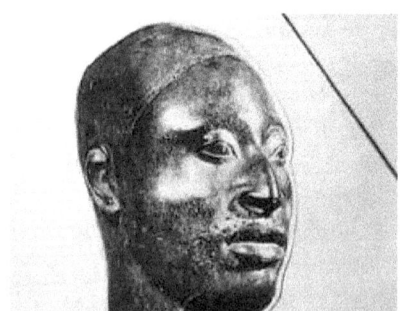

Bronze head of an ancient king from Benin, West Africa, The tradition of fine sculpture in West Africa goes back long before 1000 B.C.

Collosal head of Afro-Olmec (Xi) warrior-king, circa 1100 B.C.

Descendants of Ancient Africans in Recent America

In many parts of the Americas today, there are still people of African Negritic racial backgrounds who continue to exist either blended into the larger African-Americas population or are parts of separate, indigenous groups living on their own lands with their own unique culture and languages.

One such example is the Washitaw Nation who owned about one million square miles of the former Louisiana Territories, (see www.Hotep.org), but who now own only about 70,000 acres of all their former territory. The regaining of their lands from the U.S. was a long process which concluded partially in 1991, when they won the right to their lands in a U.S. court.

The Black Californian broke up as a nation during the late 1800's after many years of war with the Spanish invaders of the South West, with Mexico and with the U.S. The blended into the Black population of California and their descendants still exist among the millions of Black Californians of today.

The Black Caribs or Garifunas of the Caribbean Islands and Central America fought with the English and Spanish from the late fifteen hundreds up to 1797, when the British sued for peace. The Garifuna were expelled from their islands but they prospered in Central America where hundreds of thousands live along the coasts today.

The Afro-Darienite is a significant group of pre-historic, pre-columbian Blacks who existed in South America and Central America. These Blacks were the Africans that the Spanish first saw during their exploration of the narrow strip of land between Columbia and Central America and who were described as "slaves of our lord" since the

Spaniards and Europeans had the intention of enslaving all Blacks they found in the newly discovered lands.

The above mentioned Blacks of precolumbian origins are not Blacks wo mixed with the Mongoloid Indian population as occurred during the time of slavery. They were Blacks who were in some cases on their lands before the southward migrations of the Mongoloid Native Americans. In many cases, these Blacks had established civilizations in the Americas thousands of years ago.

An early Black Californian, a member of the original Black aboriginal people of California and the South Western U.S.

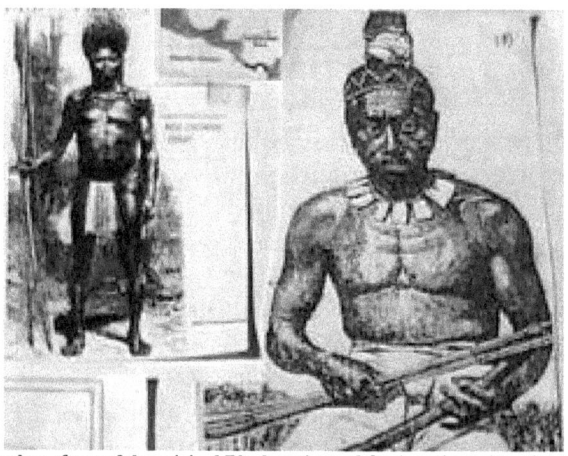

A member of one of the original Black nations of the Americas, the Afro-Darienite of Panama.

Stone carving of Negroid person found in area close to Washitaw Territories, Southern U.S.

THE USE OF ANCIENT AFRICAN SHIPS AND BOATS TO TRADE WITH THE AMERICAS

Protohistoric, prehistoric and ancient Negritic Africans were masters of the lands as well as the oceans. They were the first shipbuilders on earth and had to have used watercraft to cross from South East Asia to Australia about 60,000 years ago and from the West Africa/Sahara inland seas region to the Americas. The fact of the northern portion of Africa now known as a vast desert wasteland being a place of large lakes, rivers and fertile regions with the most ancient of civilizations is a fact that has been verified, (see African Presence in Early America, edt. Ivan Van Sertima and Runoko Rashidi, Transaction Publishers, New Bruinswick, NJ "The Principle of Polarity," by Wayne Chandler: 1994.)

From that region of Africa as well as East Africa, diffusions of Blacks towards the Americas as early as 30,000 B.C. are believed to have occurred based on findings in a region from Mexico to Brazil which show that American indians in the region include Negritic types (eg. Olmecs, Afro-Darienite, Black Californians, Chuarras, Garifunas and others). Much earlier journeys occurred by land sometime before 75,000 B.C. according to the Gladwin Thesis written by C.S. Gladwin. This migration occurred on the Pacific side of the Americas and was began by Africans with Affinities similar to the people of New Guinea, Tasmania, Solomon Islands and Australia. The earliest migrations of African Blacks through Asia then to the Americas seemed to have occurred exactly during the period that the Australian Aborigines and the proto-African ancestors of the Aborigines, Oceanic Negroids (Fijians, Solomon Islanders, Papua-New Guineans, and so on) and other Blacks spread throughout East Asia and the Pacific Islands about one hundred thousand years ago. The fact that these same Blacks are still among the world's seafaring cultures and still regard the sea as sacred and as a place of sustinence is evidence of their ancient dependance on the sea for travel and exploration as well as for commerce and trade. Therefore, they would have had to build seaworthy ships and boats to take them across the vast expanses of ocean, including the Atlantic, Indian Ocean (both the Atlantic and Indian Oceans were called the Ethiopean Sea, in the Middle Ages) and the Pacific Ocean.

During the historic period close to the early bronze or copper using period of world history (6000 B.C. to 4000 B.C. migrations of Africans from the Mende regions of West Africa and the Sahara across the Atlantic to the Americas may have occurred. In fact, the Mende agricultural culture was well established in West Africa and the Sahara during that period. Boats still criss-crossed the Sahara, as they had been doing for over ten thousand years previously. The ancient peoples of the Sahara, as rock paintings clearly show, were using boats and may have sailed from West Africa and the Sahara to the Americas, including the Washitaw territories of the Midwestern and Southern U.S. Moreover, it is believed by the aboriginal Black people of the former Washitaw Empire who still live in the Southern U.S., that about 6000 B.C., there was a great population shift from the region of Africa and the Pacific ocean, which led to the migrations of their ancestors to the Americas to join the Blacks who had been there previously.

As for the use of ships, ancient Negritic peoples and the original Negroid peoples of the earth may have began using boats very early in human history. Moreover, whatever boats were used did not have to be sophisticated or of huge size. In fact, the small, seaworthy "outrigger" canoe may have been spread from East Africa to the Indian Ocean and the Pacific by the earliest African migrants to Asia and the Pacific regions. Boats of papyrus, skin, sewed plank, log and hollowed logs were used by ancient Africans on their trips to various parts of the world.

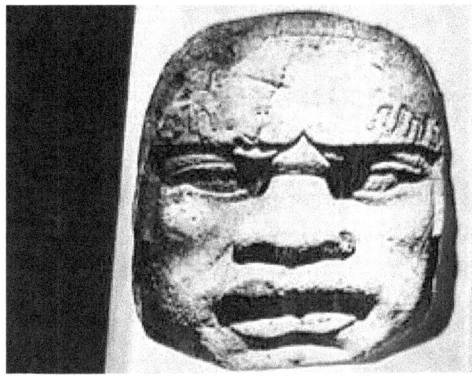
Gigantic stone head of Afro-Olmec (Xi People) of ancient Mexico, circa 1100 B.C.

Face of Afro-Olmec child carved on the waste "belt" of an Olmec ballplayer

This stone belt was used by the Olmec ballplayers to catch the impact of the rubber balls in their ball games. This face is typical Negritic, including the eyes which seem to "slant," a common racial characteristic in West Africa, the Sahara and in South Africa among the Kong-San (Bushmen) and other Africans.

TRADE ROUTES OF THE ANCIENT BLACKS

During the years of migrations of Africans to all parts of the world, those who crossed the Atlantic, Indian Ocean and Pacific also used the seas to make trips to the northern parts of Africa. They may have avoided the northern routes across the deserts at particular times of the year and sailed northward by sailing parallel to the coastslines on their way northward or southward, just as the Phoenicians, Nubians and Egyptians had done.

Boats made of skin, logs, hollowed ttee trunk, lashed canoes and skin could have been used for trading and commerce.

The reed boat is a common type of watercraft used in West Africa and other parts of the world, yet there were other boats and ships to add to those already mentioned above. Boats similar to those of Nubia and Egypt were being used in the Sahara just as long or even longer than they were being used in Egypt. In fact, civilization in the Sahara and Sudan existed before Egypt was settled by Blacks from the South and the Sahara.

The vessels which crossed the Atlantic about 1500 B.C. (during the early Afro-Olmec period) were most likely the same types of ships shown in the sahara cave paintings of ships dating to about 7,000 B.C. or similar ships from Nubian rock carvings of 3000 B.C..

Egyptologists such as Sir Flinders Petrie believed that the ancient African drawings of ships represent papyrus boats similar to the one built by the Bambara People for Thor Hayerdhal on the shores of Lake Chad. This boat made it to Barbadose, however they did not reinforce the hull with rope as the ancient Egyptians and Nubians did with their ancient ships. That lack of reinforcement made the Bambara ship weak, however another papyrus ship built by Ayamara Indians in Lake Titicaca, Bolivia was reinforced and it made it to the West Indies without difficulty.

Naval historian Bjorn Landstrom believes that some of the curved hulls shown on rock art and pottery from the Nubian civilization (circa 3000 B.C.) point to a basic three-plank idea. The planks would have been sewn together with rope. The larer version must have had some interior framing to hold them together. The hulls of some ot these boats show the vertical extension of the bow and stern which may have been to keep them bouyant.

These types of boats are stilll in use in one of the most unlikely places. The Djuka and Saramaka Tribes of Surinam, known also as 'Bush Negroes," build a style of ship and boat similar to that of the Ancient Egyptians and Nubians, with their bows and sterns curving upward and pointing vertically.

This style of boat is also a common design in parts of West Africa, particularly along the Niger River where extensive river trading occurs. They are usually carved from a single tree trunk which is used as the backbone. Planks are then fitted alongside to enlarge them. In all cases, cabins are built on top of the interior out of woven mat or other strong fiberous material. These boats are usually six to eight feet across and about fifty feet long. There is evidence that one African Emperor Abubakari of Mali used these "almadias" or longboats to make a trip to the Americas during the 1300's.(see, They Came Before Columbus, Ivan Van Sertima; Random House: 1975)

Apart from the vessels used by the West Africans and south western Sahara Black Africans to sail across the Atlantic to the Americas, Nubians, Kushites, Egyptians and Ethiopians were known traders in the Mediterranean. The Canaanites, the Negroid inhabitants of the Levant who later became the Phoenicians also were master seafarers. This has caused some to speculate that the heads of the Afro-Olmecs represent the heads of servants of the Phoenicians, yet no dominant people would build such massive and collosol monuments to their servants and not to themselves.

\ANTHROPOLOGISTS BELIEVE THERE WAS AN ANCIENT BLACK PRESENCE IN THE AMERICAS

During the International Congress of American Anthropologists held in Bacelona, Spain in 1964, a French anthropologist pointed out that all that was missing to prove a definite presence of Negritic Blacks in the Americas before Columbus was Negroid skeletons to add to the already found Negroid featured terracottas. Later on February of 1975 skeletons of Negroid people dating to the 1200's were found at a precolumbian grave in the Virgin Islands. Andrei Wierzinski, the Polish crainologist also concluded based on the study of skeletons found in Mexico, that a good portion of the skulls were that of Negritic Blacks,

Based on the many finds for a Black African Negroid presence in ancient Mexico, some of the most enthusiastic proponents of a pre-columbian Black African presence in Mexico are Mexican professionals. They conclude that Africans must have established early important trading centers on the coasts along Vera Cuz, from which Middle America's first civiliztion grew.

In retrospect, ancient Africans did visit the Americas from as early as about 100,000 B.C. where they stayed for tens of thousands of years. By 30,000 B.C., to about 15,000 B.C., a massive migration from the Sahara towards the Indian Ocean and the Pacific in the East occurred from the Sahara. Blacks also migrated Westward across the Atlantic Ocean towards the Americas during that period until the very eve of Columbus' first journey to the Americas.

Trade, commerce and exploration as well as the search for new lands when the Sahara began to dry up later in history was the catalyst that drove the West Africans towards the Atlantic and into the Americas.

REFERENCES

Washitaw Nation (Hotep Organization)

Clyde A. Winters (*The Nubians and the Olmecs*)

Blacks of India (Dalitstan Organization)

Blacks of the Pacific and Melanesia:
www.cwo.com/~lucumi/pacific.html

DESCENDANTS OF PRECOLUMBIAN BLACKS IN THE U.S., CARIBBEAN, CENTRAL AMERICA AND SOUTH AMERICA AND THE FIGHT FOR THE RETURN OF THEIR STOLEN OCCUPIED LANDS IN THE MIDST OF THE REPARATIONS DEBATE THE ISSUE OF RETURNING THE LANDS OF THESE BLACKS WHO ANCESTORS WERE HERE IN THE U.S. AND AMERICAS BEFORE COLUMBUS HAS ALREADY BEEN DONE WITH ONE BLACK NATION OF THE LOUISIANA TERRITORIES

The experience of the Washitaw Nation (or Ouchita Nation) of the Southern United States is another piece of solid evidence for the fact of pre-Columbian African presence and settlement in the Americas and specifically in the United States. According to an article carried in the magazine, 'The Freedom Press Newsletter, (Spring, 1996), reprinted from Earthways, The Newsletter of the Sojourner Truth Farm School (August, 1995), the Washitaw were (and still are) a nation of Africans who existed in the Southern U.S. and Mississippi Valley region long before the 16th century Europeans arrived and even before there were "Native Americans" on the lands the Washitaw once occupied and still occupy today.

According to the article, "the Washitaw Nation "governed three million acres of land in Louisiana, Arkansas, Oklahoma, Texas and Mississippi. They were ship builders (similar to the Garifuna of the Caribbean, who are also of pre-Columbian West Afrucan Mandinka Muslim origins (according to Harold Lawrence in 'African Presence in Early America,edt. by Ivan Van Sertima).

What is even more facinating about this aspect of hidden history of Blacks in America before Columbus is that the Washitaw Nation was known and recognized as a separate, independant Black nation by the Spanish and French, who were in the Louisiana Territories and Texas areas. According to the present leader of the Washitaw Nation, "when Spain ceded the Louisiana Territory to France, they excluded the land belonging to the Washitaw Nation. France did not include it in the "Louisiana Purchase," and according to the leader, "This land is not part of the United States of America." That point was made in the newspaper, "The Capitol Spotlight, June 1992.

In fact, the courts agreed that the land was not part of the U.S. and that in fact the Washitaw (Ouchita) Nation was on the land long before European Colonization: therefore, in legal decisions made, some of the *ancient territory was returned. This historical decision was made about 1991.*

This is the type of information seldom seen in the majority press, yet, the importance of that event clearly points to the incredible service small papers and magazines such as Ancient American or the Capitol Spotlight and The Freedom Press Newsletter have been making, along iwth internet news and information sites such as this one. So, here we see an example in the continental United States where Africans who came before slavery, before Columbus and thousands of years before Christ (over six thousand years B.C., according to the Washitaw chroniclers), were engaged in boat building, seafaring, trade and commerce in ancient times and who still exist today as a distinct Black Nation who have evidence and proof of their ownership of millions of acres of lands in the Southern U.S. and the Mississippi Valley. The Washitaw Nation held an important convention in June 1992, in Monroe, Louisiana and have held others since. (see www.Hotep.org for the Washitaw's point of view on their history and culture).

Yet, the Washitaw is merely one nation of the descendants of pre-columbian Blacks from Africa and elsewhere and possibly from right here in the Americas as the very first people to exist here, long before the development of the Mongoloid, American Indians or the Mongoloid(15,000 B.C.) or even the Caucasian races (30,000 B.C.). Pure Black Homosapiens began to migrate from Africa and populate the entire earth about 200,000 to 150,000 years ago, according to scientists, historians and anthropologists.

Among the other Black nations who existed in the Americas before Columbus and long before Christ were the Jamassee (Yamassee), who had a large kingdom in the South eastern U.S., Their descendants were among the first Blacks of pre-columbian American origins who fell victim to kidnapping for the purpose of enslavement. Blacks of South America, the Caribbean and Central America were also attacked and enslaved based on a Pontifax passed during the mid- 1400's by the Church hierachy giving the Europeans the go ahead to enslave all "Children of Ham" found in the newly discovered territories. The descendants of the Jamassee are the millions of Blacks who live in Alabama, Gerogia, South Carolina and northern Florida. They of course also have African slave ancestors, but these slaves are the relatives of the same Africans who sailed to America of their own free will, while Europe was in the Dark Ages, and long before Christ, for that matter.

In California, descendnats of the fierce "Black Californians" who were a Negroid people of African racial origins and the original owners of California and the South WEST (BEFORE THE SPANISH INVSION...OR THE CREATION OF THE MIXED RACE "HISPANIC" ETHNIC GROUP.

Many African-Americans in California are of Black Californian ancestry and their great grand parents were among the original Black Californians who were victims of Spanish Californio enslavement and Anglo American settler attacks. In fact, the Black Californian fought until the late 1800's to maintain control of their ancestral lands from the settlers. THAT'S A FACT.

There are aboriginal nations of Blacks in Panama such as the Afro-Darienite and the Choco people. In fact, the Afro-Darienite are the remnants of the aboriginal Black nations of South and Central America who were once hunted down to be made slaves by the Spaniards (in fact Balboa or Peter Matyr chroniclers referred to these Blacks as "slaves of our lord,") meaning, like Blacks in Africa, the South Pacific and elsewhere, they were eligible for enslavement, being descended from Ham, the so-called "father of the Black race."

In Columbia's Choco Region, on the Western side of that country, there are hundreds of thousands of Blacks, whose ancestors have been in Columbia for thousands of years. In fact, scientists and some historians have found out that Black slaves were being kidnapped and hunted down in Columbia and parts of South and Central America, as well as the Caribbean and U.S., by the Spaniards and others long before they began to look for slaves in Africa. (an old painting in Natonal Geographic clearly shows a black with bow and arrow and wearing a loin cloth, hunting along the coast of Columbia during the first voyage there by the Spaniards. These Blacks today of the Choco Region of Columbia are among the most oppressed of Blacks in Latin America today (See the Final Call back issues on this topic)

Then there is the Garifuna or Kalifunami also called "Black Caribs" Being a member of the Black Carib Nation and having done historical research, the myth of the Black Caribs being escaped slaves has been debunked. It is true that the Black Caribs encouraged slaves from the West Indies Islands to join them and that the Black Caribs did ally with the Mongoloid Caribs of Dominica and other parts of the West Indies, but the fact remains, that the Black Caribs were originally Mende traders of gold and cloth, who established settlements throughout the Circum-Caribbean region, Mexico, Central America, South America and the Southern U.S. They had been arriving in the Americas for thousands of years, even before they converted to Islam during the 900's A.D.. In fact, the

Olmecs of ancient Mexico were Mende, they used the Mende script (found on monments at Monte Alban, Mexico, and they named places from southern Mexico to South America with Mandinka names. Such names sometimes sound identical to the names of places used in West Africa.

In retrospect, while the debate for reparations increses, it is important that African-Americans know that two great injustices were committed by the Europeans. The first was slavery, the second was the taking of Black lands and destroying Black history and culture so Blacks remain totally ignorant of their rights to more than one third of north America. NOW YOU KNOW WHY THE SLAVEMASTERS DID NOT WANT BLACK FOLK TO LEARN TO READ, AND WHY PLANTS ARE PLACED IN CHATROOMS AND ON FORUMS TO ATTEMPT TO DISCREDIT ANY USEFUL HISTORY AND INFORMATION OFFERED TO BLACK PEOPLE.

Still, TRUTH SUBMERGED SHALL RISE AGAIN.

SUSU ECONOMICS

THE HISTORY OF PAN-AFRICAN TRADE, COMMERE, WEALTH AND MONEY

(A Preview of the Facinating History of the Development of Ancient Black Civilizations Worldwide)

One of the most important aspects of Black history worldwide is the development of Black civilization due to the early and persisten use and application of trade and commerce. Due to such early and well organized trading and commercial systems throughout the prehistoric Black world, Blacks were able to expand throughout the world and establish the world's first cultures and civilizations. Although it is said that Blacks migrated from the original homeland of mankind in Africa to settle all Asia, Europe, Australia and the Americas (see Scientific American; Sept. 2000, p. 80-87...this is a recent publication), long before the differentiation of the races from the original Negritic to Negriic, Caucasoid, Mongoloid, along with the various mixed races such as Polynesians, Native Americans, Japanese, Malays, Mediterranean whites, East Indians (the mixed Negroid/Caucasian type...not the pure Black pre Aryan Negritic Indians), Arabs, Latinos (Mestizos, Mullatoes, Zambos, Spaniards) and a number of other mixed races and regional types, the purpose of the earlies migrations of Blacks from Africa to the rest of the world was not merely following and hunting wild animals, as some theorists have claimed, but searching for commodities, like red ocre to paint the smooth, dark skin from insects and decoration. Another purpose for the early migrations of Africans to other parts of the world was to establish trading and commercial links to those of their own people, who had left previously. Hence, even if the earliest migrations were after wandering herds of animals, further migrations were in search of links with their kinsmen and women.

The migrations of Africans to all parts of the world within the past hundred thousand years or more occurred before other races existed. Thus, Black culture and civilization was being established when no other "races" existed as we know them today. This is a facinating historical even, because having been homosapiens for over one hundred thousand years, it is very possible that Blacks could have gone through many periods of cultural development and civilization before the beginning of the Nile Valley civilization (since about 17,000 B.C.) or the Zingh Civilization of the South-Western Sahara (15,000 B.C.), or even Atlantis (10,000 B.C.), or the building of the Sphinx (7,000 B.C.).

In fact, there is evidence from ancient East Indian chronicles (some of these pictures are on AAWR (African American Web Ring) of the geat scientific advancement of the Black prehistoric inhabitants of the Indus Valley Civilization (6000 b.c. to 1700 b.c), who built flying machines, who had flushing toilets, cities on a grid-like pattern, and many of what we may call "modern" conveniences.

About 20,000 years ago, the present-day dried up and desertified Sahara had an aquatic civilization where the Africans who lived on the edges of the giant inland sea, built large ocean-going ships. Rock paintings of these ships can still be seen in the Sahara (and some appeared on national geographic magazine about two years ago). (For more on the Aquatic Civilizations of the prehistoric Sahara, see, "African Presence In Early Asia," by Ivan Van Sertima and Runoko Rashidi, Transaction Publications, New Bruinswick, NJ).

The Africans who used these boats (which are still used today by tribes such as the Baduma of Mali, West Africa) made of papyrus straw. These same type of boats were used to travel to the Americas, the Indian Ocean, the South Pacific, India, East Asia and the Pacific, then to the Americas via the Pacific Ocean. In fact, the Fijians still consider Africa's East Coast to be their very ancient homeland and Africans in East Africa have oral as well as written histories of ancient journeys towards Asia.

In ancient times, trade between Africans in Africa and those in the Indian Ocean, East Asia and the Pacific Ocean, East Asia, the Americas, the Mediterranean, the Black Sea area and all the continents including Australia. In all these areas, evidence of prehistoric African Blacks exist. IT IS VERY IMPORTANT TO NOTE THAT SUCH EVIDENCE WAS AGAIN FOUND IN SOUTH AMERICA, WHERE ABOUT FIFTY SKULLS REPRESENTING NEGROID PEOPLE WERE FOUND IN BRAZIL (see Scientific American, September 2000). However, this is no news to some Blacks, particularly those descended from the ancient prehistoric Blacks of America, such as the Wasitaw of the Louisiana area, the descendants of the Black Californians, the Jamassee and others; the Black Caribs of the Caribbean and Central America, the Choco Region Blacks of Columbia, South America and many others.

We have contributed to the development of Russia (i.e., Alexander Pushkin). Through Ivan Sertima's book *They Came Before Columbus*, we see that Africans were on the soil of the Americas before the arrival of Columbus.

What happened?

We brought some of the destruction upon ourselves. Augustine (St. Augustine, or St. Austin) the Carthaginian (originally from Hippo Regius) brought in the Vandals to fight off enemies of Carthage (but the Vandals destroyed everything; hence the word vandalism). We told others of our wealth, such as Leo Africanus (1494 – 1554?). We displayed our wealth, as did Mansa Musa (ruled 1312 - 1337). We kept knowledge too confined. Other powers acquired certain of our brainpower to further their development, while destroying others of our traditional sources of brainpower and practices. Images of foreign powers were projected upon us, while our own positive images were destroyed. Names of significant places and institutions were changed to reflect foreign ideas and conquests. We were made to study the positive value of foreign powers, while belittling our own. We have accepted words like "minority" and "disadvantaged". They call their leaders of tribes Kings and call our leaders of tribes Chiefs. A King reflects more global significance than a Chief. Everything Britain did was in the name of the King or Queen. Look at Russia today. Just a few years ago, they were competing for one of the world's top three or four super powers. With the change of defining their system of operations from communism to 'free market' economy, Russia is in a mental tailspin. What happened, the people are the same, but the way they now see themselves has changed. When you are in transition, you must be careful to not let define your directions.

Regardless of seeming overbearing obstacles, we still rose. Throughout international slavery and the slave trade, we still produced Black people who contributed significantly to the development of the world's civilizations. In science and technology, we produced thousands of brilliant minds such as Benjamin Banneker, Granville T. Woods, Lewis Latimer, Garrett Morgan, George Washington Carver, Percy Julian, Daniel hale Williams, Charles Drew, Jan E. Matzeliger, etc. To help project images of these contributors, the African Scientific Institute's *Blacks In Science Calendar* was produced from 1986 through 1996. Authors have published articles and books about our contributions in science and technology.

Past

(0 AD – 1800)

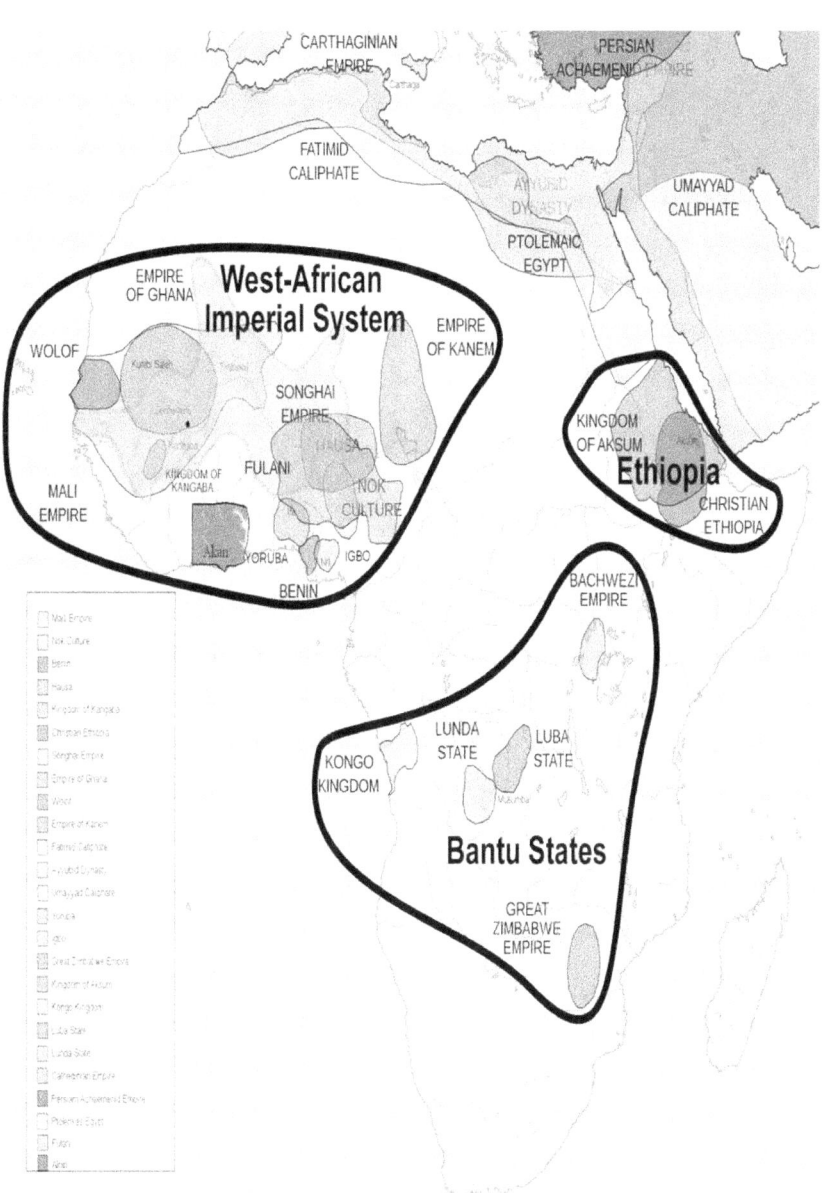

96

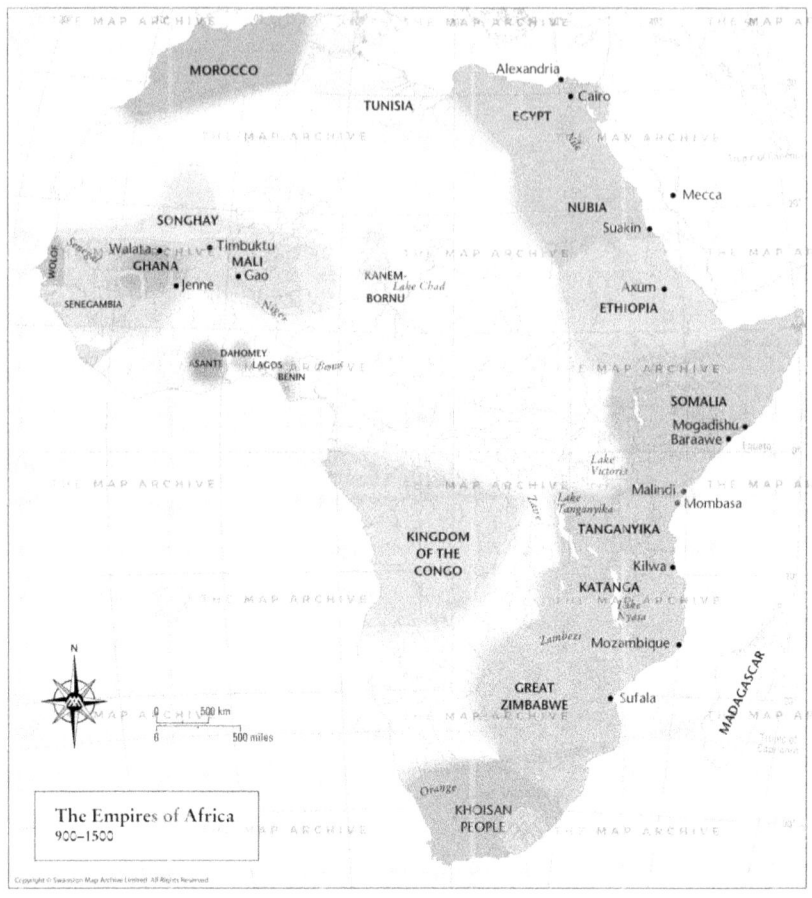

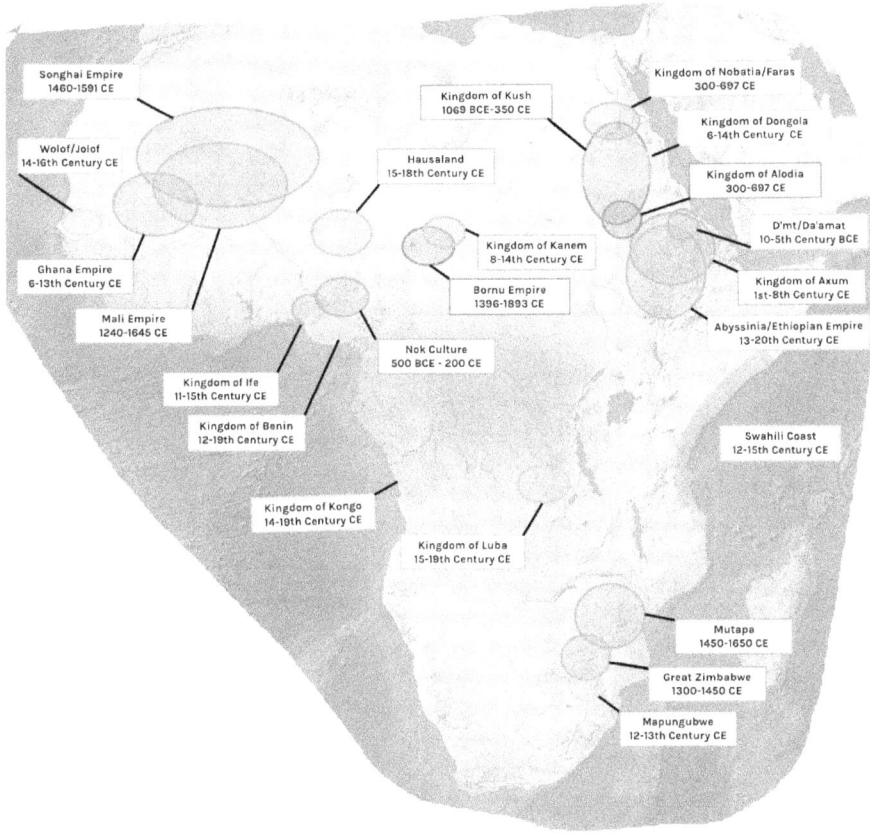

African Empires

There were several civilizations in Africa from 50 AD – 1700. We are only listing a few of them,

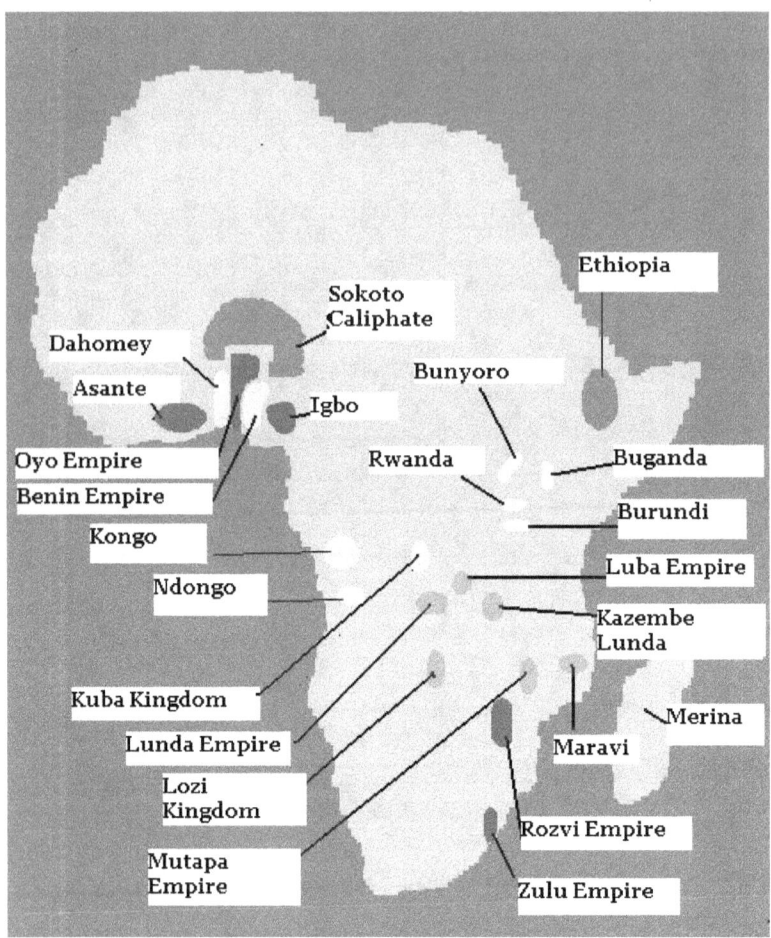

Image for Afropedea

African Kingdoms

Also, some areas of Africa were controlled by Kingdoms (not vast areas some people would consider Empires)

- Kingdom of Bazin (9th century)
- Kingdom of Belgin (9th century)
- Kingdom of Jarin (9th century)
- Kingdom of Qita'a (9th century)
- Kingdom of Nagash (9th century)
- Kingdom of Tankish (9th century)
- Sultanate of Mogadishu (10th century – 16th century)

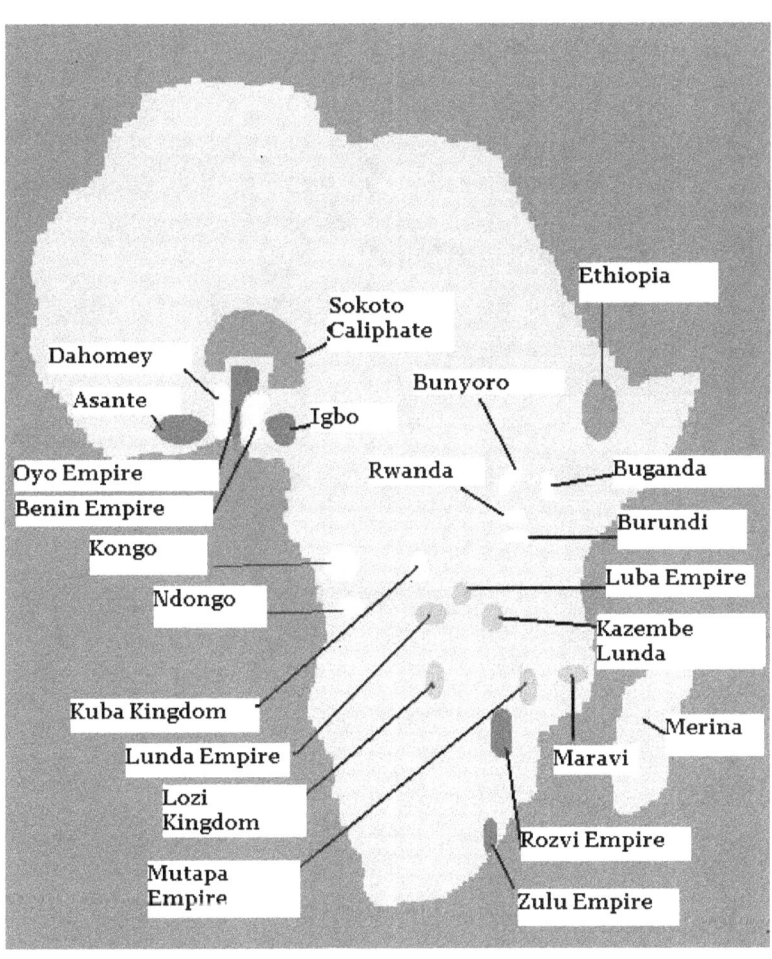

Image for Afropedea

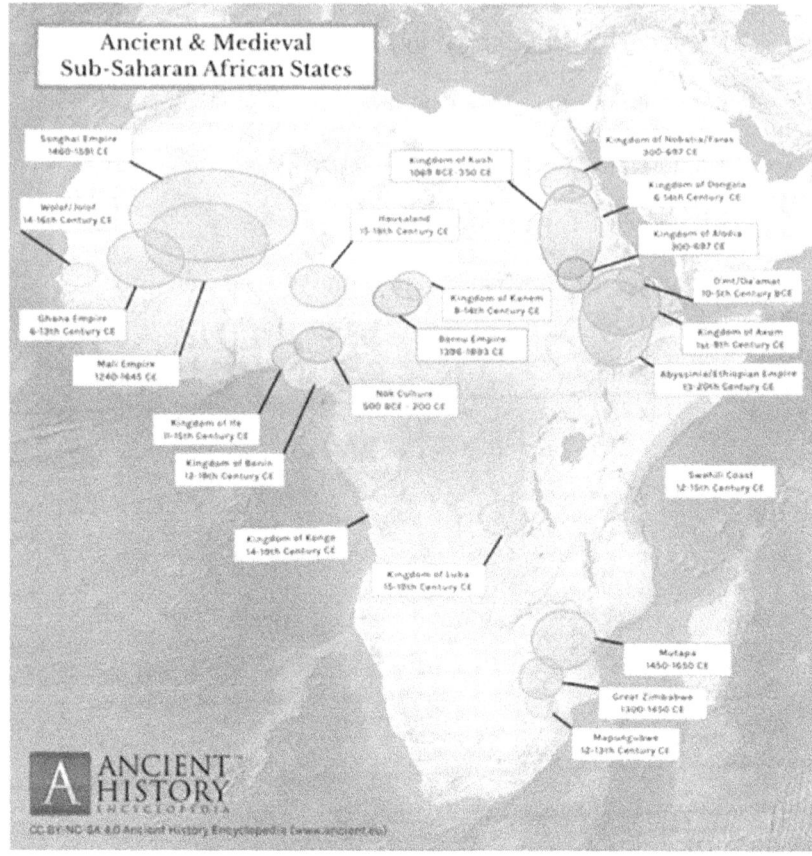

****

Book by Robin Walker

Opinion by *By Robin Walker © 2006*

THIS IS AN EXCTRACT FROM A BOOK BY ROBIN WALKER

1. **The human race is of African origin.** The oldest known skeletal remains of anatomically modern humans (or *homo sapiens*) were excavated at sites in East Africa. Human remains were discovered at Omo in Ethiopia that were dated at 195,000 years old, the oldest known in the world.

2. **Skeletons of pre-humans have been found in Africa** that date back between 4 and 5 million years. The oldest known ancestral type of humanity is thought to have been the *australopithecus ramidus*, who lived at least 4.4 million years ago

3. **Africans were the first to organise fishing expeditions** 90,000 years ago. At Katanda, a region in northeastern Zaïre (now Congo),

was recovered a finely wrought series of harpoon points, all elaborately polished and barbed. Also uncovered was a tool, equally well crafted, believed to be a dagger. The discoveries suggested the existence of an early aquatic or fishing based culture.

4. **Africans were the first to engage in mining** 43,000 years ago. In 1964 a hematite mine was found in Swaziland at Bomvu Ridge in the Ngwenya mountain range. Ultimately 300,000 artefacts were recovered including thousands of stone-made mining tools. Adrian Boshier, one of the archaeologists on the site, dated the mine to a staggering 43,200 years old.

5. **Africans pioneered basic arithmetic** 25,000 years ago. The Ishango bone is a tool handle with notches carved into it found in the Ishango region of Zaïre (now called Congo) near Lake Edward. The bone tool was originally thought to have been over 8,000 years old, but a more sensitive recent dating has given dates of 25,000 years old. On the tool are 3 rows of notches. Row 1 shows three notches carved next to six, four carved next to eight, ten carved next to two fives and finally a seven. The 3 and 6, 4 and 8, and 10 and 5, represent the process of doubling. Row 2 shows eleven notches carved next to twenty-one notches, and nineteen notches carved next to nine notches. This represents 10 + 1, 20 + 1, 20 - 1 and 10 - 1. Finally, Row 3 shows eleven notches, thirteen notches, seventeen notches and nineteen notches. 11, 13, 17 and 19 are the prime numbers between 10 and 20.

6. **Africans cultivated crops 12,000 years ago**, the first known advances in agriculture. Professor Fred Wendorf discovered that people in Egypt's Western Desert cultivated crops of barley, capers, chick-peas, dates, legumes, lentils and wheat. Their ancient tools were also recovered. There were grindstones, milling stones, cutting blades, hide scrapers, engraving burins, and mortars and pestles.

7. **Africans mummified their dead 9,000 years ago**. A mummified infant was found under the Uan Muhuggiag rock shelter in south western Libya. The infant was buried in the foetal position and was mummified using a very sophisticated technique that must have taken hundreds of years to evolve. The technique predates the earliest mummies known in Ancient Egypt by at least 1,000 years. Carbon dating is controversial but the mummy may date from 7438 (±220) BC.

8. **Africans carved the world's first colossal sculpture** 7,000 or more

years ago. The Great Sphinx of Giza was fashioned with the head of a man combined with the body of a lion. A key and important question raised by this monument was: How old is it? In October 1991 Professor Robert Schoch, a geologist from Boston University, demonstrated that the Sphinx was sculpted between 5000 BC and 7000 BC, dates that he considered conservative.

9. **On the 1 March 1979, the** *New York Times* carried an article on its front page also page sixteen that was entitled *Nubian Monarchy called Oldest*. In this article we were assured that: "Evidence of the oldest recognizable monarchy in human history, preceding the rise of the earliest Egyptian kings by several generations, has been discovered in artifacts from ancient Nubia" (i.e. the territory of the northern Sudan and the southern portion of modern Egypt.)

10. **The ancient Egyptians had the same type** of tropically adapted skeletal proportions as modern Black Africans. A 2003 paper appeared in *American Journal of Physical Anthropology* by Dr Sonia Zakrzewski entitled *Variation in Ancient Egyptian Stature and Body Proportions* where she states that: "The raw values in Table 6 suggest that Egyptians had the 'super-Negroid' body plan described by Robins (1983). The values for the brachial and crural indices show that the distal segments of each limb are longer relative to the proximal segments than in many 'African' populations."

11. **The ancient Egyptians had Afro combs.** One writer tells us that the Egyptians "manufactured a very striking range of combs in ivory: the shape of these is distinctly African and is like the combs used even today by Africans and those of African descent."

12. **The Funerary Complex in the ancient Egyptian** city of Saqqara is the oldest building that tourists regularly visit today. An outer wall, now mostly in ruins, surrounded the whole structure. Through the entrance are a series of columns, the first stone-built columns known to historians. The North House also has ornamental columns built into the walls that have papyrus-like capitals. Also inside the complex is the Ceremonial Court, made of limestone blocks that have been quarried and then shaped. In the centre of the complex is the Step Pyramid, the first of 90 Egyptian pyramids.

13. **The first Great Pyramid of Giza,** the most extraordinary building in history, was a staggering 481 feet tall - the equivalent of a 40-storey building. It was made of 2.3 million blocks of limestone and granite, some weighing 100 tons.

14. **The ancient Egyptian city of Kahun** was the world's first planned city. Rectangular and walled, the city was divided into two parts. One part housed the wealthier inhabitants – the scribes, officials and foremen. The other part housed the ordinary people. The streets of the western section in particular, were straight, laid out on a grid, and crossed each other at right angles. A stone gutter, over half a metre wide, ran down the centre of every street.

15. **Egyptian mansions were discovered in Kahun** - each boasting 70 rooms, divided into four sections or quarters. There was a master's quarter, quarters for women and servants, quarters for offices and finally, quarters for granaries, each facing a central courtyard. The master's quarters had an open court with a stone water tank for bathing. Surrounding this was a colonnade.

16 **The Labyrinth in the Egyptian city of Hawara** with its massive layout, multiple courtyards, chambers and halls, was the very largest building in antiquity. Boasting three thousand rooms, 1,500 of them were above ground and the other 1,500 were underground.

17. **Toilets and sewerage systems** existed in ancient Egypt. One of the pharaohs built a city now known as Amarna. An American urban planner noted that: "Great importance was attached to cleanliness in Amarna as in other Egyptian cities. Toilets and sewers were in use to dispose waste. Soap was made for washing the body. Perfumes and essences were popular against body odour. A solution of natron was used to keep insects from houses . . . Amarna may have been the first planned 'garden city'."

18. **Sudan has more pyramids than any other country on earth** - even more than Egypt. There are at least 223 pyramids in the Sudanese cities of Al Kurru, Nuri, Gebel Barkal and Meroë. They are generally 20 to 30 metres high and steep sided.

19. **The Sudanese city of Meroë** is rich in surviving monuments. Becoming the capital of the Kushite Empire between 590 BC until AD 350, there are 84 pyramids in this city alone, many built with their own

miniature temple. In addition, there are ruins of a bath house sharing affinities with those of the Romans. Its central feature is a large pool approached by a flight of steps with waterspouts decorated with lion heads.

20. **Bling culture has a long and interesting history.** Gold was used to decorate ancient Sudanese temples. One writer reported that: "Recent excavations at Meroe and Mussawwarat es-Sufra revealed temples with walls and statues covered with gold leaf".

21. **In around 300 BC, the Sudanese invented a writing script** that had twenty-three letters of which four were vowels and there was also a word divider. Hundreds of ancient texts have survived that were in this script. Some are on display in the British Museum.

22. **In central Nigeria, West Africa's oldest civilisation** flourished between 1000 BC and 300 BC. Discovered in 1928, the ancient culture was called the Nok Civilisation, named after the village in which the early artefacts were discovered. Two modern scholars, declare that "[a]fter calibration, the period of Nok art spans from 1000 BC until 300 BC". The site itself is much older going back as early as 4580 or 4290 BC.

23. **West Africans built in stone by 1100 BC.** In the Tichitt-Walata region of Mauritania, archaeologists have found "large stone masonry villages" that date back to 1100 BC. The villages consisted of roughly circular compounds connected by "well-defined streets".

24. **By 250 BC, the foundations of West Africa's oldest cities were established** such as Old Djenné in Mali

25. **Kumbi Saleh, the capital of Ancient Ghana,** flourished from 300 to 1240 AD. Located in modern day Mauritania, archaeological excavations have revealed houses, almost habitable today, for want of renovation and several storeys high. They had underground rooms, staircases and connecting halls. Some had nine rooms. One part of the city alone is estimated to have housed 30,000 people.

26. **West Africa had walled towns and cities** in the pre-colonial period. Winwood Reade, an English historian visited West Africa in the nineteenth century and commented that: "There are . . . thousands of

large walled cities resembling those of Europe in the Middle Ages, or of ancient Greece.

27. **Lord Lugard, an English official, estimated in 1904** that there were 170 walled towns still in existence in the whole of just the Kano province of northern Nigeria.

28. **Cheques are not quite as new an invention** as we were led to believe. In the tenth century, an Arab geographer, Ibn Haukal, visited a fringe region of Ancient Ghana. Writing in 951 AD, he told of a cheque for 42,000 golden dinars written to a merchant in the city of Audoghast by his partner in Sidjilmessa.

29. **Ibn Haukal, writing in 951 AD,** informs us that the King of Ghana was "the richest king on the face of the earth" whose pre-eminence was due to the quantity of gold nuggets that had been amassed by the himself and by his predecessors.

30. **The Nigerian city of Ile-Ife was paved in 1000 AD** on the orders of a female ruler with decorations that originated in Ancient America. Naturally, no-one wants to explain how this took place approximately 500 years before the time of Christopher Columbus!

31. **West Africa had bling culture in 1067 AD.** One source mentions that when the Emperor of Ghana gives audience to his people: "he sits in a pavilion around which stand his horses caparisoned in cloth of gold: behind him stand ten pages holding shields and gold-mounted swords: and on his right hand are the sons of the princes of his empire, splendidly clad and with gold plaited into their hair . . . The gate of the chamber is guarded by dogs of an excellent breed . . . they wear collars of gold and silver."

32. **Glass windows existed at that time.** The residence of the Ghanaian Emperor in 1116 AD was: "A well-built castle, thoroughly fortified, decorated inside with sculptures and pictures, and having glass windows."

33. **The Grand Mosque in the Malian city of Djenné,** described as "the largest adobe [clay] building in the world", was first raised in 1204 AD. It was built on a square plan where each side is 56 metres in length. It has three large towers on one side, each with projecting wooden buttresses.

34. **One of the great achievements of the Yoruba** was their urban culture. "By the year A.D. 1300," says a modern scholar, "the Yoruba people built numerous walled cities surrounded by farms". The cities were Owu, Oyo, Ijebu, Ijesa, Ketu, Popo, Egba, Sabe, Dassa, Egbado, Igbomina, the sixteen Ekiti principalities, Owo and Ondo.

35. **Yoruba metal art of the mediaeval period** was of world class. One scholar wrote that Yoruba art "would stand comparison with anything which Ancient Egypt, Classical Greece and Rome, or Renaissance Europe had to offer."

36. **In the Malian city of Gao** stands the Mausoleum of Askia the Great, a weird sixteenth century edifice that resembles a step pyramid.

37. **Thousands of mediaeval tumuli** have been found across West Africa. Nearly 7,000 were discovered in north-west Senegal alone spread over nearly 1,500 sites. They were probably built between 1000 and 1300 AD.

38. **Excavations at the Malian city of Gao** carried out by Cambridge University revealed glass windows. One of the finds was entitled: "Fragments of alabaster window surrounds and a piece of pink window glass, Gao 10th – 14th century."

39. **In 1999 the BBC produced a television series entitled** *Millennium.* The programme devoted to the fourteenth century opens with the following disclosure: "In the fourteenth century, the century of the scythe, natural disasters threatened civilisations with extinction. The Black Death kills more people in Europe, Asia and North Africa than any catastrophe has before. Civilisations which avoid the plague thrive. In West Africa the Empire of Mali becomes the richest in the world."

40. **Malian sailors got to America in 1311 AD,** 181 years before Columbus. An Egyptian scholar, Ibn Fadl Al-Umari, published on this sometime around 1342. In the tenth chapter of his book, there is an account of two large maritime voyages ordered by the predecessor of Mansa Musa, a king who inherited the Malian throne in 1312. This mariner king is not named by Al-Umari, but modern writers identify him as Mansa Abubakari II.

41. **On a pilgrimage to Mecca in 1324 AD,** a Malian ruler, Mansa Musa, brought so much money with him that his visit resulted in the collapse of gold prices in Egypt and Arabia. It took twelve years for the economies of the region to normalise.

42. **West African gold mining took place on a vast scale.** One modern writer said that: "It is estimated that the total amount of gold mined in West Africa up to 1500 was 3,500 tons, worth more than $30 billion in today's market."

43. **The old Malian capital of Niani** had a 14th century building called the Hall of Audience. It was an surmounted by a dome, adorned with arabesques of striking colours. The windows of an upper floor were plated with wood and framed in silver; those of a lower floor were plated with wood, framed in gold.

44. **Mali in the 14th century was highly urbanised.** Sergio Domian, an Italian art and architecture scholar, wrote the following about this period: "Thus was laid the foundation of an urban civilisation. At the height of its power, Mali had at least 400 cities, and the interior of the Niger Delta was very densely populated".

45. **The Malian city of Timbuktu** had a 14th century population of 115,000 - 5 times larger than mediaeval London. Mansa Musa, built the Djinguerebere Mosque in the fourteenth century. There was the University Mosque in which 25,000 students studied and the Oratory of Sidi Yayia. There were over 150 Koran schools in which 20,000 children were instructed. London, by contrast, had a total 14th century population of 20,000 people.

46. *National Geographic* **recently described Timbuktu** as the Paris of the mediaeval world, on account of its intellectual culture. According to Professor Henry Louis Gates, 25,000 university students studied there.

47. **Many old West African families have private library collections** that go back hundreds of years. The Mauritanian cities of Chinguetti and Oudane have a total of 3,450 hand written mediaeval books. There may be another 6,000 books still surviving in the other city of Walata. Some date back to the 8th century AD. There are 11,000 books in

private collections in Niger. Finally, in Timbuktu, Mali, there are about 700,000 surviving books.

48. **A collection of one thousand six hundred books** was considered a small library for a West African scholar of the 16th century. Professor Ahmed Baba of Timbuktu is recorded as saying that he had the smallest library of any of his friends - he had only 1600 volumes.

49. **Concerning these old manuscripts,** Michael Palin, in his TV series Sahara, said the imam of Timbuktu "has a collection of scientific texts that clearly show the planets circling the sun. They date back hundreds of years . . . Its convincing evidence that the scholars of Timbuktu knew a lot more than their counterparts in Europe. In the fifteenth century in Timbuktu the mathematicians knew about the rotation of the planets, knew about the details of the eclipse, they knew things which we had to wait for 150 almost 200 years to know in Europe when Galileo and Copernicus came up with these same calculations and were given a very hard time for it."

50. **The Songhai Empire of 16th century West Africa** had a government position called Minister for Etiquette and Protocol.

51. **The mediaeval Nigerian city of Benin** was built to "a scale comparable with the Great Wall of China". There was a vast system of defensive walling totalling 10,000 miles in all. Even before the full extent of the city walling had become apparent the *Guinness Book of Records* carried an entry in the 1974 edition that described the city as: "The largest earthworks in the world carried out prior to the mechanical era."

52. **Benin art of the Middle Ages** was of the highest quality. An official of the Berlin Museum für Völkerkunde once stated that: "These works from Benin are equal to the very finest examples of European casting technique. Benvenuto Cellini could not have cast them better, nor could anyone else before or after him . . . Technically, these bronzes represent the very highest possible achievement."

53. **Winwood Reade described his visit to the Ashanti Royal Palace** of Kumasi in 1874: "We went to the king's palace, which consists of many courtyards, each surrounded with alcoves and verandahs, and having two gates or doors, so that each yard was a thoroughfare . . . But the part of the palace fronting the street was a stone house, Moorish in its style . . . with a flat roof and a parapet, and suites of apartments on

the first floor. It was built by Fanti masons many years ago. The rooms upstairs remind me of Wardour Street. Each was a perfect Old Curiosity Shop. Books in many languages, Bohemian glass, clocks, silver plate, old furniture, Persian rugs, Kidderminster carpets, pictures and engravings, numberless chests and coffers. A sword bearing the inscription *From Queen Victoria to the King of Ashantee*. A copy of the *Times,* 17 October 1843. With these were many specimens of Moorish and Ashanti handicraft."

54. **In the mid-nineteenth century, William Clarke,** an English visitor to Nigeria, remarked that: "As good an article of cloth can be woven by the Yoruba weavers as by any people . . . in durability, their cloths far excel the prints and home-spuns of Manchester."

55. **The recently discovered 9th century Nigerian city of Eredo** was found to be surrounded by a wall that was 100 miles long and seventy feet high in places. The internal area was a staggering 400 square miles.

56. **On the subject of cloth,** Kongolese textiles were also distinguished. Various European writers of the sixteenth and seventeenth centuries wrote of the delicate crafts of the peoples living in eastern Kongo and adjacent regions who manufactured damasks, sarcenets, satins, taffeta, cloth of tissue and velvet. Professor DeGraft-Johnson made the curious observation that: "Their brocades, both high and low, were far more valuable than the Italian."

57. **On Kongolese metallurgy of the Middle Ages,** one modern scholar wrote that: "There is no doubting . . . the existence of an expert metallurgical art in the ancient Kongo . . . The Bakongo were aware of the toxicity of lead vapours. They devised preventative and curative methods, both pharmacological (massive doses of pawpaw and palm oil) and mechanical (exerting of pressure to free the digestive tract), for combating lead poisoning."

58. **In Nigeria, the royal palace** in the city of Kano dates back to the fifteenth century. Begun by Muhammad Rumfa (ruled 1463-99) it has gradually evolved over generations into a very imposing complex. A colonial report of the city from 1902, described it as "a network of buildings covering an area of 33 acres and surrounded by a wall 20 to 30 feet high outside and 15 feet inside . . . in itself no mean citadel".

59. **A sixteenth century traveller** visited the central African civilisation of Kanem-Borno and commented that the emperor's cavalry had golden "stirrups, spurs, bits and buckles." Even the ruler's dogs had "chains of the finest gold".

60. **One of the government positions** in mediaeval Kanem-Borno was Astronomer Royal.

61. **Ngazargamu, the capital city of Kanem-Borno**, became one of the largest cities in the seventeenth century world. By 1658 AD, the metropolis, according to an architectural scholar housed "about quarter of a million people". It had 660 streets. Many were wide and unbending, reflective of town planning.

62. **The Nigerian city of Surame** flourished in the sixteenth century. Even in ruin it was an impressive sight, built on a horizontal vertical grid. A modern scholar describes it thus: "The walls of Surame are about 10 miles in circumference and include many large bastions or walled suburbs running out at right angles to the main wall. The large compound at Kanta is still visible in the centre, with ruins of many buildings, one of which is said to have been two-storied. The striking feature of the walls and whole ruins is the extensive use of stone and *tsokuwa* (laterite gravel) or very hard red building mud, evidently brought from a distance. There is a big mound of this near the north gate about 8 feet in height. The walls show regular courses of masonry to a height of 20 feet and more in several places. The best preserved portion is that known as sirati (the bridge) a little north of the eastern gate . . . The main city walls here appear to have provided a very strongly guarded entrance about 30 feet wide."

63. **The Nigerian city of Kano** in 1851 produced an estimated 10 million pairs of sandals and 5 million hides each year for export.

64. **In 1246 AD Dunama II of Kanem-Borno** exchanged embassies with Al-Mustansir, the king of Tunis. He sent the North African court a costly present, which apparently included a giraffe. An old chronicle noted that the rare animal "created a sensation in Tunis".

65. **By the third century BC the city of Carthage** on the coast of Tunisia was opulent and impressive. It had a population of 700,000 and

may even have approached a million. Lining both sides of three streets were rows of tall houses six stories high.

66. The Ethiopian city of Aksum has a series of 7 giant obelisks that date from perhaps 300 BC to 300 AD. They have details carved into them that represent windows and doorways of several stories. The largest obelisk, now fallen, is in fact "the largest monolith ever made anywhere in the world". It is 108 feet long, weighs a staggering 500 tons, and represents a thirteen-storey building

67. Ethiopia minted its own coins over 1,500 years ago. One scholar wrote that: "Almost no other contemporary state anywhere in the world could issue in gold, a statement of sovereignty achieved only by Rome, Persia, and the Kushan Kingdom in northern India at the time."

68. The Ethiopian script of the 4th century AD influenced the writing script of Armenia. A Russian historian noted that: "Soon after its creation, the Ethiopic vocalised script began to influence the scripts of Armenia and Georgia. D. A. Olderogge suggested that Mesrop Mashtotz used the vocalised Ethiopic script when he invented the Armenian alphabet.

69. "In the first half of the first millennium CE," says a modern scholar, Ethiopia "was ranked as one of the world's greatest empires". A Persian cleric of the third century AD identified it as the third most important state in the world after Persia and Rome.

70. Ethiopia has 11 underground mediaeval churches built by being carved out of the ground. In the twelfth and thirteenth centuries AD, Roha became the new capital of the Ethiopians. Conceived as a New Jerusalem by its founder, Emperor Lalibela (c.1150-1230), it contains 11 churches, all carved out of the rock of the mountains by hammer and chisel. All of the temples were carved to a depth of 11 metres or so below ground level. The largest is the House of the Redeemer, a staggering 33.7 metres long, 23.7 metres wide and 11.5 metres deep.

71. Lalibela is not the only place in Ethiopia to have such wonders. A cotemporary archaeologist reports research that was conducted in the region in the early 1970's when: "startling numbers of churches built in caves or partially or completely cut from the living rock were revealed not only in Tigre and Lalibela but as far south as Addis Ababa. Soon at least 1,500 were known. At least as many more probably await revelation."

72. **In 1209 AD Emperor Lalibela of Ethiopia** sent an embassy to Cairo bringing the sultan unusual gifts including an elephant, a hyena, a zebra, and a giraffe.

73. **In Southern Africa, there are at least 600 stone built ruins** in the regions of Zimbabwe, Mozambique and South Africa. These ruins are called Mazimbabwe in Shona, the Bantu language of the builders, and means great revered house and "signifies court".

74. **The Great Zimbabwe was the largest of these ruins.** It consists of 12 clusters of buildings, spread over 3 square miles. Its outer walls were made from 100,000 tons of granite bricks. In the fourteenth century, the city housed 18,000 people, comparable in size to that of London of the same period.

75. **Bling culture existed in this region.** At the time of our last visit, the Horniman Museum in London had exhibits of headrests with the caption: "Headrests have been used in Africa since the time of the Egyptian pharaohs. Remains of some headrests, once covered in gold foil, have been found in the ruins of Great Zimbabwe and burial sites like Mapungubwe dating to the twelfth century after Christ."

76. **Dr Albert Churchward, author of *Signs and Symbols of Primordial Man*,** pointed out that writing was found in one of the stone built ruins: "Lt.-Col. E. L. de Cordes . . . who was in South Africa for three years, informed the writer that in one of the 'Ruins' there is a 'stone-chamber,' with a vast quantity of Papyri, covered with old Egyptian hieroglyphics. A Boer hunter discovered this, and a large quantity was used to light a fire with, and yet still a larger quantity remained there now."

77. **On bling culture,** one seventeenth century visitor to southern African empire of Monomotapa, that ruled over this vast region, wrote that: "The people dress in various ways: at court of the Kings their grandees wear cloths of rich silk, damask, satin, gold and silk cloth; these are three widths of satin, each width four covados [2.64m], each sewn to the next, sometimes with gold lace in between, trimmed on two sides, like a carpet, with a gold and silk fringe, sewn in place with a two fingers' wide ribbon, woven with gold roses on silk."

78. **Southern Africans mined gold on an epic scale.** One modern writer tells us that: "The estimated amount of gold ore mined from the entire region by the ancients was staggering, exceeding 43 million tons. The ore yielded nearly 700 tons of pure gold which today would be valued at over $7.5 billion."

79. **Apparently the Monomotapan royal palace at Mount Fura** had chandeliers hanging from the ceiling. An eighteenth century geography book provided the following data: "The inside consists of a great variety of sumptuous apartments, spacious and lofty halls, all adorned with a magnificent cotton tapestry, the manufacture of the country. The floors, cielings [sic], beams and rafters are all either gilt or plated with gold curiously wrought, as are also the chairs of state, tables, benches &c. The candle-sticks and branches are made of ivory inlaid with gold, and hang from the cieling by chains of the same metal, or of silver gilt."

80. **Monomotapa had a social welfare system.** Antonio Bocarro, a Portuguese contemporary, informs us that the Emperor: "shows great charity to the blind and maimed, for these are called the king's poor, and have land and revenues for their subsistence, and when they wish to pass through the kingdoms, wherever they come food and drinks are given to them at the public cost as long as they remain there, and when they leave that place to go to another they are provided with what is necessary for their journey, and a guide, and some one to carry their wallet to the next village. In every place where they come there is the same obligation."

81. **Many southern Africans have indigenous and pre-colonial words for 'gun'.** Scholars have generally been reluctant to investigate or explain this fact.

82. **Evidence discovered in 1978 showed that East Africans** were making steel for more than 1,500 years: "Assistant Professor of Anthropology Peter Schmidt and Professor of Engineering Donald H. Avery have found as long as 2,000 years ago Africans living on the western shores of Lake Victoria had produced carbon steel in preheated forced draft furnaces, a method that was technologically more sophisticated than any developed in Europe until the mid-nineteenth century."

83. **Ruins of a 300 BC astronomical observatory** was found at Namoratunga in Kenya. Africans were mapping the movements of stars such as Triangulum, Aldebaran, Bellatrix, Central Orion, etcetera, as well as the moon, in order to create a lunar calendar of 354 days.

84. **Autopsies and caesarean operations** were routinely and effectively carried out by surgeons in pre-colonial Uganda. The surgeons routinely used antiseptics, anaesthetics and cautery iron. Commenting on a Ugandan caesarean operation that appeared in the *Edinburgh Medical Journal* in 1884, one author wrote: "The whole conduct of the operation . . . suggests a skilled long-practiced surgical team at work conducting a well-tried and familiar operation with smooth efficiency."

85. **Sudan in the mediaeval period** had churches, cathedrals, monasteries and castles. Their ruins still exist today.

86. **The mediaeval Nubian Kingdoms kept archives.** From the site of Qasr Ibrim legal texts, documents and correspondence were discovered. An archaeologist informs us that: "On the site are preserved thousands of documents in Meroitic, Latin, Greek, Coptic, Old Nubian, Arabic and Turkish."

87. **Glass windows existed in mediaeval Sudan.** Archaeologists found evidence of window glass at the Sudanese cities of Old Dongola and Hambukol.

88. **Bling culture existed in the mediaeval Sudan.** Archaeologists found an individual buried at the Monastery of the Holy Trinity in the city of Old Dongola. He was clad in an extremely elaborate garb consisting of costly textiles of various fabrics including gold thread. At the city of Soba East, there were individuals buried in fine clothing, including items with golden thread.

89. **Style and fashion existed in mediaeval Sudan.** A dignitary at Jebel Adda in the late thirteenth century AD was interned with a long coat of red and yellow patterned damask folded over his body. Underneath, he wore plain cotton trousers of long and baggy cut. A pair of red leather slippers with turned up toes lay at the foot of the coffin. The body was wrapped in enormous pieces of gold brocaded striped silk.

90. **Sudan in the ninth century AD** had housing complexes with bath rooms and piped water. An archaeologist wrote that Old Dongola, the capital of Makuria, had: "a[n] . . . eighth to . . . ninth century housing complex. The houses discovered here differ in their hitherto unencountered spatial layout as well as their functional programme (water supply installation, bathroom with heating system) and interiors decorated with murals."

91. **In 619 AD, the Nubians** sent a gift of a giraffe to the Persians.

92. **The East Coast, from Somalia to Mozambique, has ruins of well over 50 towns** and cities. They flourished from the ninth to the sixteenth centuries AD.

93. **Chinese records of the fifteenth century AD** note that Mogadishu had houses of "four or five stories high".

94. **Gedi, near the coast of Kenya,** is one of the East African ghost towns. Its ruins, dating from the fourteenth or fifteenth centuries, include the city walls, the palace, private houses, the Great Mosque, seven smaller mosques, and three pillar tombs.

95. **The ruined mosque in the Kenyan city of Gedi** had a water purifier made of limestone for recycling water.

96. **The palace in the Kenyan city of Gedi** contains evidence of piped water controlled by taps. In addition it had bathrooms and indoor toilets.

97. **A visitor in 1331 AD considered the Tanzanian city** of Kilwa to be of world class. He wrote that it was the "principal city on the coast the greater part of whose inhabitants are Zanj of very black complexion." Later on he says that: "Kilwa is one of the most beautiful and well-constructed cities in the world. The whole of it is elegantly built."

98. **Bling culture existed in early Tanzania.** A Portuguese chronicler of the sixteenth century wrote that: "[T]hey are finely clad in many rich garments of gold and silk and cotton, and the women as well; also with much gold and silver chains and bracelets, which they wear on their legs and arms, and many jewelled earrings in their ears".

99. **In 1961 a British archaeologist,** found the ruins of Husuni Kubwa, the royal palace of the Tanzanian city of Kilwa. It had over a hundred rooms, including a reception hall, galleries, courtyards, terraces and an octagonal swimming pool.

100. **In 1414 the Kenyan city of Malindi sent ambassadors to China** carrying a gift (Giraffe) that created a sensation at the Imperial Court.

****

Northern Africa

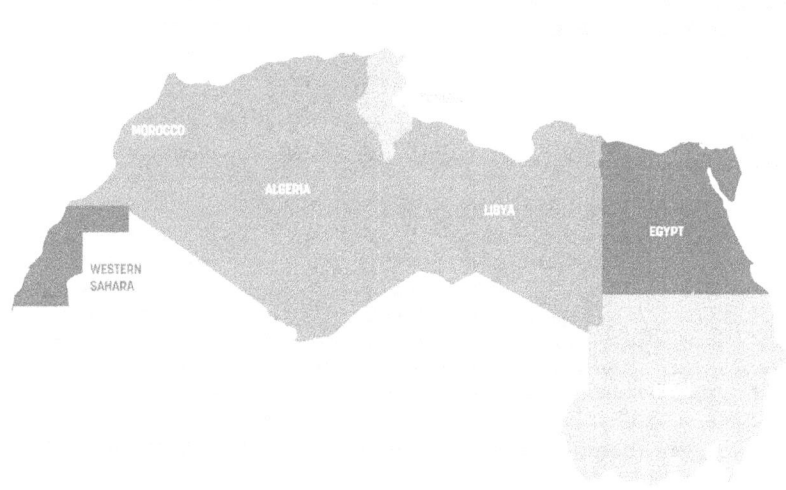

from Wikipedia

Islamic North African empires
- In Algeria:
    - Banu Ifran dynasty (830–1040)
    - Zirid dynasty (947–1090)
    - Hammadid dynasty (1014–1152)
    - Almohad caliphate (1145–1244)
- In Morocco:
    - Idrisid dynasty (789–974)
    - Almoravid dynasty (1061–1145)
    - Almohad dynasty (1145–1244)
    - Marinid dynasty (1244–1465)
    - Wattasid dynasty (1471–1554)
    - Saadi dynasty (1554-1666)
    - Alaouite dynasty (1666–present)
- In Tunisia:
    - Fatimid dynasty (Tunisian period) (910–969)
    - Zirid dynasty (973–1148)
    - Hafsid dynasty (1229–1574)
- In Egypt:

- Fatimid dynasty (Egyptian period) (969–1171)
- Ayyubid dynasty (1171–1254)
- Mamluk dynasty (1250–1517)

- In Sudan:
    - The Sennar Sultanate (1502–1821) was a sultanate in the north of Sudan. It was named Funj after the ethnic group of its dynasty or Sinnar (or Sennar) after its capital, which ruled a substantial area of the Sudan region.

Almoravid Empire (1056 – 1147)

The Almoravid Empire stretched from the Ebro valley to today's Mauritania. Its administration was organised centrally, with Almoravid dignitaries at its head. Members of the Malikite judiciary were in charge of religious and judicial affairs. The empire's mints were supplied with African gold, with which they produced Dinars, which found their way to the Christian kingdoms of Spain (They were given the name marabotins).

Not much has been written concerning Almoravid urban construction yet they were responsible for two of today's major urban centres in Morocco: Marrakech and Fes. Almoravid art brought a number of innovations to the architectural tradition of the Muslim west. The use of multi-foiled cusped arches was already known in Umayyad al-Andalus.

Almoravid ghazi

Western Africa

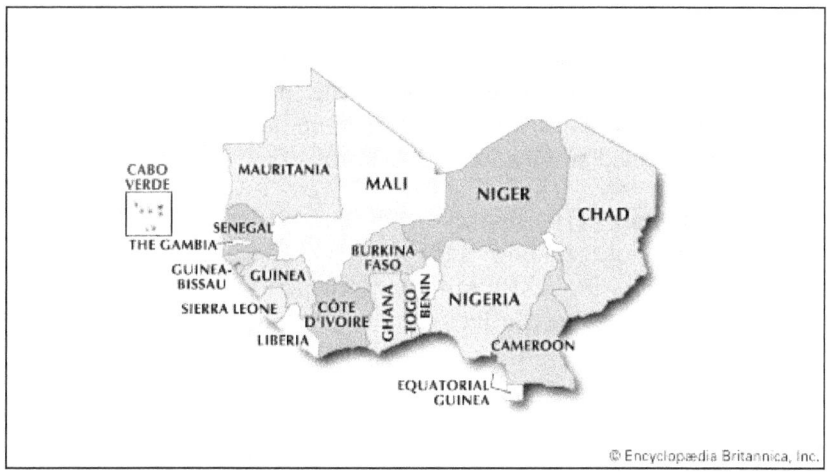

- The **Kingdom of Nri** (1043–1911) was the West African state of the Nri-Igbo, a subgroup of the Igbo people, and is the oldest kingdom in Nigeria.

- The **Oyo Empire** (1400–1895) was a powerful Yoruba state in the modern country of Nigeria. It began in the 1300s in the West African savannah north of the tropical forests where other Yoruba peoples lived. Being in the savannah proved beneficial, as Oyo could use horses, which were unable to live farther south (due to the tsetse fly). Using this armored cavalry, the empire was able to extend its reach across parts of what is now northern and western Nigeria. Oyo was generally unable to penetrate to the coast except where the savanna reached southward to the ocean in Benin. There, Oyo forced the kingdoms of Allada and Dahomey into subordination which gave the empire access to European trade. (from http://slaveryandremembrance.org/articles/article/?id=A0121)

- **Benin Empire** (1240–1897), a pre-colonial African empire of modern Nigeria. The empire once stretched to present day Ghana ruled by sky kings (OGISO) in the first dynasty and by OBAS in the second dynasty. It was the first kingdom to come in contact with the Europeans.. The **Kingdom of Benin**, also known as the **Benin Kingdom**, was a pre-colonial kingdom in what is now southern Nigeria. It is not to be confused with Benin, the post-colonial nation state. The Kingdom of Benin's capital was Edo, now known as Benin City in Edo state. The Benin Kingdom was "one of the oldest and most highly developed states in the coastal hinterland of West Africa", it was formed around the 11th century.

- **Bonoman** (11th century–19th century), earliest known Akan state. Gold trading and Kola nut trading with Northern Neighbors brought wealth and prosperity to Akan creators of this state. Culture influenced much of modern Akan culture.

****

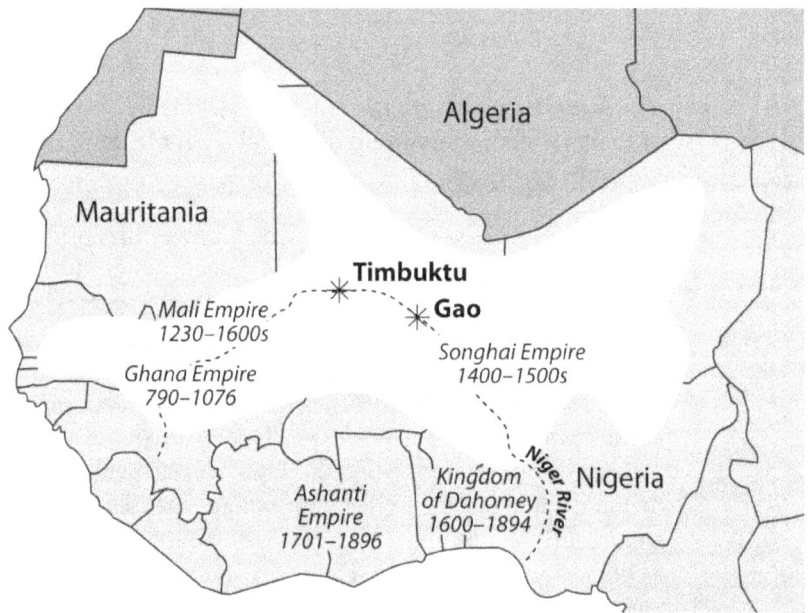

Ghana Empire

It was the first to benefit from the introduction of gold mining. Ghana dominated the region between about 750 and 1078.

From African Glory, by J.C. deGraft-Johnson:

Ghana's commercial relations with the outside world were very important. It lay at the southern end of the western caravan route that ran from Sijilmasa in modern Morocco on through Taghaza in the Sahara desert, famous throughout recorded history for its salt mines. The principle imports of the Ghana Empire were clothes, brocades, copper, and salt, while the chief item exported was Ghana's valuable gold dust.

According to entrees in the *Dictionary of Black Africa Civilization*, by Georges Balandier and Jacques Maquet, Arab and Sudanese chroniclers never tired of describing the riches and power of the King who could raise an army of 200,000 men. Although gold mining was an occupation of the forest lands and not under the direct control of the King, the sovereigns monopolized its export of gold. The same was true of salt, which was extracted by Saharan Berbers.

Mali Empire (1240-1645 CE)

The Mali Empire was founded by Sundiata Keita (r. 1230-1255 CE) following his victory over the kingdom of Sosso (c. 1180-1235 CE). Sundiata's centralised government, diplomacy and well-trained army permitted a massive military expansion which would pave the way for a flourishing of the Mali Empire, making it the largest yet seen in Africa.

The reign of Mansa Musa I (1312-1337 CE) saw the empire reach new heights in terms of territory controlled, cultural fluorescence, and the staggering wealth brought through Mali's control of regional trade routes. Acting as a middle-trader between North Africa via the Sahara desert and the Niger River to the south, Mali exploited the traffic in gold, salt, copper, ivory, and slaves that crisscrossed West Africa. Muslim merchants were attracted to all this commercial activity, and they converted Mali rulers who in turn spread Islam via such noted centres of learning as Timbuktu. In contrast to cities like Niani (the capital), Djenne, and Gao, most of the rural Mali population remained farmers who clung to their traditional animist beliefs. The Mali Empire collapsed in the 1460s CE following civil wars, the opening up of trade routes elsewhere, and the rise of the neighboring Songhai Empire, but it did continue to control a small part of the western empire into the 17th century CE.

Bibliography

- Desmond Clark, J. (ed). *The Cambridge History of Africa, Volume 1.* Cambridge University Press, 2001.
- Garlake, P. *Early Art and Architecture of Africa.* Oxford University Press, 2002.
- Hrbek, I. (ed). *UNESCO General History of Africa, Vol. III, Abridged Edition.* University of California Press, 1992.
- Ki-Zerbo, J. (ed). *UNESCO General History of Africa, Vol. IV, Abridged Edition.* University of California Press, 1998.
- Oliver, R. (ed). *The Cambridge History of Africa, Vol. 3.* Cambridge University Press, 2001.
- Curtin, P. *African History.* Pearson, 1995.

Ashanti Empire (1701–1894)

The empire stretched from central Ghana to present day Togo and Côte d'Ivoire, bordered by the **Dagomba kingdom to** the north and Dahomey to the east.

The Ashanti Empire was a powerful kingdom located in the area of present-day Ghana on the coast of West Africa and centered on the capital of Kumasi. Because of its vast deposits of gold, which were highly sought after by European traders, this area became known as the Gold Coast. During the eighteenth and nineteenth centuries Ashanti was the largest and most powerful of the states formed by the Akan peoples of West Africa.

By the end of the century the Ashanti Empire was at its peak, ruling three to four million people and controlling nearly five hundred miles of coastline. By the end of the century the Ashanti Empire was at its peak, ruling three to four million people and controlling nearly five hundred miles of coastline.

During this time European entrepreneurs were establishing trading posts along the Gold Coast. The Ashanti took advantage of this opportunity to sell their valuable resources of gold and slaves in exchange for firearms and gunpowder, which they used to bolster their military power. The Ashanti had long employed slave labor for their own use, but now they found a ready and lucrative export market among the Europeans. Indeed, the Ashanti became an important supplier of African laborers to the transatlantic slave trade.

Songhai Empire (1375 – 1591)

The Songhai Empire was the largest and last of the three major pre-colonial empires to emerge in West Africa. From its capital at Gao on the Niger River, Songhai expanded in all directions until it stretched from the Atlantic Ocean (modern Senegal and Gambia) to what is now Northwest Nigeria and central Niger. Gao, Songhai's capital, which remains to this day a small Niger River trading center, was home to the famous Goa Mosque and the Tomb of Askia, the most important of the Songhai emperors. The cities of Timbuktu and Djenne were the other major cultural and commercial centers of the empire. (from BlackPast.org, https://www.blackpast.org/global-african-history/songhai-empire-ca-1375-1591/)

Economic Structure

Safe economic trade existed throughout the Empire, due to the 200,000 person army stationed in the provinces. Primary to the economic foundation of the Songhai Empire were the gold fields of the Niger River. These gold fields, which were often independently operated, provided a steady supply of gold that could be purchased and bartered for salt. Salt was considered so precious a commodity in West Africa that it was not uncommon for gold to be traded for equal weight in salt. When coupled with the sale of slaves, salt and gold consisted of the bulk of trans-Saharan trade and the Songhai dominance in these commodities solidified Songhai's role as a leader in the trans-Saharan trade system.

Empire of Kanem

The Kanem-Bornu Empire was a large African state which existed from the 9th century through the end of the 19th century and which spanned a region which today includes the modern-day countries of Niger, Chad, Cameroon, and Nigeria. The empire was founded by the Zaghawa nomadic people, who may have been the first in the central Sudan to acquire and make use of iron technology and horses. (BlackPast.org, https://www.blackpast.org/global-african-history/empire-kanem-bornu-c-9th-century-1900/

The Kingdom of Kanem (aka Kanim) was an ancient African state located in modern-day Chad, which flourished from the 9th to 14th century CE. With its heartland in the centre of the African continent on the eastern shores of Lake Chad, the kingdom was formed by a confederation of nomadic peoples and then ruled by the Saifawa dynasty. The city prospered thanks to its position as the hub of trade connections with central African peoples, the Nile Valley, and North African states on the other side of the Sahara Desert. The kingdom adopted the Islamic religion after long contact with Muslim clerics and traders from the 11th century CE onwards. In the 1390s CE Kanem's was forced to flee the invading Bulala people and so set up a new state on the other side of Lake Chad, which would become the Bornu Empire, sometimes known as the Kanem-Bornu Empire, which lasted until the late 19th century CE.

Central Africa

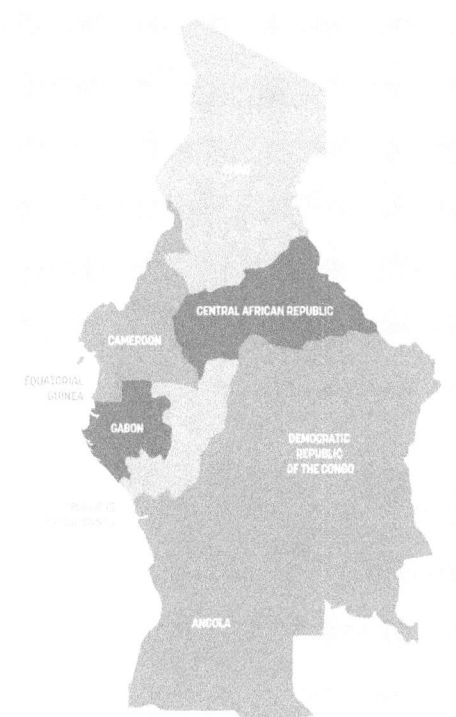

The **Kongo Kingdom** was a large kingdom in the western part of central Africa. The name comes from the fact that the founders of the kingdom were KiKongo speaking people, and the spelling of Congo with a C comes from the Portuguese translation. Kingdom was founded around 1390 CE through the political marriage of Nima a Nzima, of the Mpemba Kasi, and Luqueni Luansanze, of the Mbata, which cemented the alliance between the two KiKongo speaking peoples. The Kingdom would reach its peak in the mid 1600s. The Kingdom of Kongo would eventually fall to scheming nobles, feuding royal factions, and the Trans-Atlantic slave trade, initiating its eventual decline. (from South African History Online (SAHO), https://www.sahistory.org.za/article/kingdom-kongo-1390-1914

KONGO IN 1648

The Kingdom of Kongo, with a population of well over 2 million people at its peak, prospered thanks to trade in ivory, copper, salt, cattle hides, and slaves. The latter trade was especially lucrative and well-regulated, with rotating markets appearing in towns on fixed days of the week selling slaves acquired from the upper reaches of the Congo River. In addition to acquiring goods from elsewhere, the kingdom produced its own goods via specialised groups of craftworkers such as weavers (who produced the famous raffia fabrics of Kongo), potters, and metalworkers.

The **Luba Empire** (1585–1885) arose in the marshy grasslands of the Upemba Depression in what is now southern Democratic Republic of Congo.

The origins of the Kingdom of Luba rose in central Africa around 1300 CE in the southern rain forests of the Shaba, spreading to cover the wet grasslands of the Lake Upemba Depression (in the southern part of today's Democratic Republic of the Congo, formerly known as Zaire). The Luba kings and oral traditions claimed a past connection with the Shaba area inhabited by the Iron Age Lualaba people. Lualaba sites, the earliest being Kamilamba, date to the 6-7th century CE, and they thrived thanks to local metal deposits, especially copper, which allowed them to trade with other central African peoples using the region's rivers.

Lunda Empire (1660–1887) in what is now the Democratic Republic of Congo, north-eastern Angola and northwestern Zambia. Its central state was in Katanga.

This was an empire in Central Africa, located along the Kisai tributaries, which was very influential in the modern day Shaba Province of the Democratic Republic of Congo, eastern Angola, and Western Zambia.

The empire grew by controlling the lucrative trade route from the Kwango River to Kisai tributaries. The capital would later be moved to the Kisai tributaries to exploit trade. It collected tribute from conquered people in foodstuffs like yams, millet, cassava, and maize. The empire exchanged foodstuff for copper, salt, raffia cloth, and tobacco. The kingdom initially did not have a large population for widescale farming, so it began to raid villages and polities for such endeavors. Later raiding would be practice for the more lucrative Atlantic slave trade. The Lunda would eventually sell its criminals into slavery instead of executing them. It is estimated that 1/3 of slave coming from the region was originating from the Lunda Empire, after 1850. The empire also profited from the ivory trade, which did not last due to diminished elephant population, in the latter part of the 1800s.

Eastern Africa

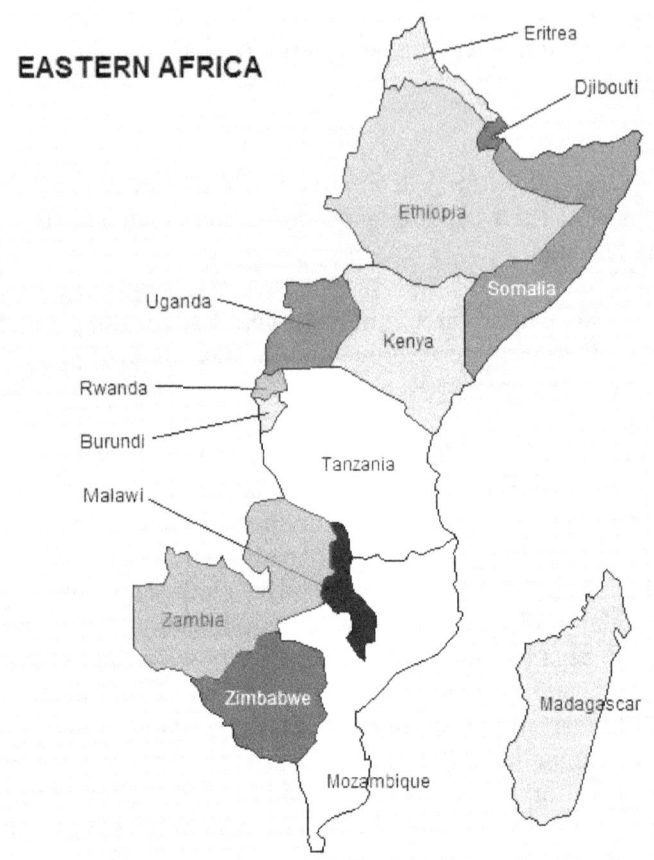

An **Empire of Kitara** in the area of the African Great Lakes has long been treated as a historical entity, but is now mostly considered an unhistorical narrative created as a response to the dawn of rule under the **Lwo Empire**.

According to Bunyoro-Kitara Kingdom, General Information (http://www.bunyoro-

kitara.org/resources/Bunyoro+Kitara+Kingdom$2C+General+Informat
ion.pdf

In the oral tradition, Kitara was a kingdom which, at the height of its power in the fourteenth and fifteenth ...centuries, included much of Uganda, northern Tanzania and eastern Congo (DRC), ruled by a dynasty known as the Bachwezi (or Chwezi) who were the successors of the Batembuzi Dynasty.According to the story, the Kitara Empire lasted until the 16th century, when it was invaded by Luo people, who came from the South of the present-day Sudan and established the kingdom of Bunyoro-Kitara. Evidence suggests that the clans of Buganda, for instance, have their own history (based on oral tradition) that is exclusive of the history of the Kingdom of Buganda

The Bachwezi are credited with the founding of the ancient empire of Kitara; which included areas of present day central, western, and southern Uganda; northern Tanzania, western Kenya, and eastern Congo. Very little is documented about them. Their entire reign was shrouded in mystery, so much so that they were accorded the status of demi gods and worshipped by various clans. Many traditional gods in Toro, Bunyoro and Buganda have typical kichwezi (adjective) names like Ndahura, Mulindwa, Wamara, Kagoro, etc..

The **Buganda** (1300–present)

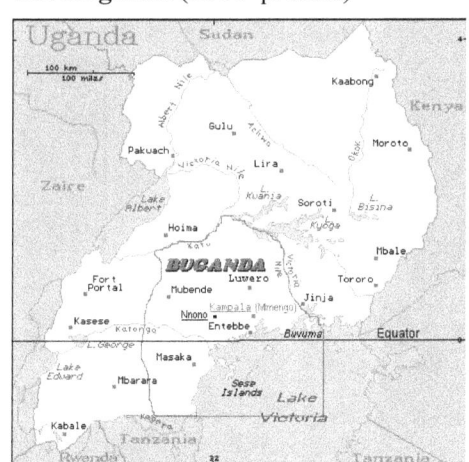

The Kingdom of Buganda, from which modern Uganda derives its name, is one of the oldest traditional kingdoms in East Africa, with a history that dates back some 1,000 years.

Today, the kingdom covers all of Uganda's Central Region, including capital Kampala, while the kingdom's subjects account for more than 20 percent of Uganda's total population of 34.9 million.

The Baganda, who number about 3 million people, are located along the northern and western shores of Lake Victoria in the East African nation of Uganda. The former Kingdom of *Buganda* is bounded on the north by the former Kingdom of Bunyoro, on the east by the Nile River, and on the west by the former kingdoms of Ankole and Toro. To the south of Buganda is the present country of Tanzania. The Baganda are the largest tribe in Uganda, and the Kingdom of Buganda is the largest of the former kingdoms.

Swahili city-states

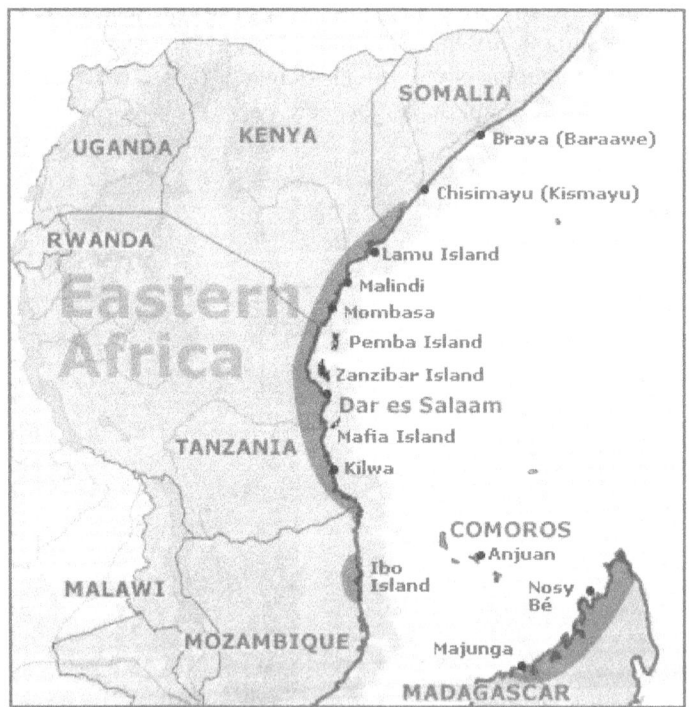

From approximately 1000 to 1500 AD, a number of city-states on the eastern coast of Africa participated in an international trade network and became cosmopolitan Islamic cultural centers. The major autonomous, but symbiotic, city-states stretched over 1,500 miles from Mogadishu (in modern day Somalia) in the north to Sofala (in modern Mozambique) in the south and included Mombasa, Gedi, Pate, Lamu, Malindi, Zanzibar, and Kilwa.

These city-states also exported natural resources. Local merchants gathered ivory from the south, gold from the western interior and frankincense and myrrh from northern Africa. Kilwa, Pate, and Mogadishu also developed a local textile industry while Kilwa and Mogadishu extracted copper from nearby mines. All of the states produced pottery. Ironworking had evolved in East Africa before the rise of the city states. They improved the process and produced iron objects for trade as well as local use.

Kingdom of Aksum (1st century – 9th century)

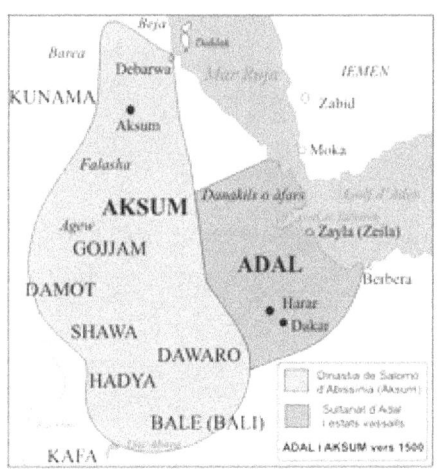

Aksum developed a civilization and empire whose influence, at its height in the 4th and 5th centuries C.E., extended throughout the regions lying south of the Roman Empire, from the fringes of the Sahara in the west, across the Red Sea to the inner Arabian desert in the east. The Aksumites developed Africa's only indigenous written script, Ge'ez. They traded with Egypt, the eastern Mediterranean and Arabia.

Despite its power and reputation—it was described by a Persian writer as one of the four greatest powers in the world at the time—very little is known about Aksum

Sultanates of East Africa

Sultanate of Ifat (1285–1415)

A Muslim Sultanate in the Horn of Africa. Led by the Walashma dynasty, it was centered in the ancient cities of Zeila and Shewa. The Kingdom ruled over parts of what are now eastern Ethiopia, Djibouti, and northern Somalia. Ifat first emerged in the 13th century, when Sultan Umar Walashma conquered the Sultanate of Showa in 1285. The Adal Sultanate or Kingdom of Adal was founded after the fall of the Sultanate of Ifat. It flourished from around 1415 to 1577.

The Sultanate of Mogadishu was an important trading empire that lasted from the 10th century to the 16th century. It maintained a vast trading network, dominated the regional gold trade, minted its own Mogadishu currency, and left an extensive architectural legacy in present-day southern Somalia.

The **Ajuran Sultanate** ruled over large parts of the Horn of Africa between the 13th and late 17th centuries. The Ajuran Sultanate is an imamate that emerged in south-central Somalia from the mid-13th century, a theocratic Islamic state based on Marka or Merca about 75 miles south of Mogadishu. The imamate was subdivided into regions administered by amirs (governors). Ajuran developed efficient communications as well as a disciplined military system, including a standing army of infantrymen accompanied by cavalry. One factor in Ajuran's effectiveness was its use of terror tactics as an instrument of warfare. The Ajuran are remembered for their public works: many of the deep stone-lined wells and fortifications still standing in southern Somalia are attributed to Ajuran engineering skills. (from Wiley Online Library:https://onlinelibrary.wiley.com/doi/abs/10.1002/9781118455074.wbeoe146)

The **Warsangali Sultanate** was a kingdom centered in northeastern and in some parts of southeastern Somalia. It was one of the largest sultanates ever established in the territory.

Bunyoro Empire

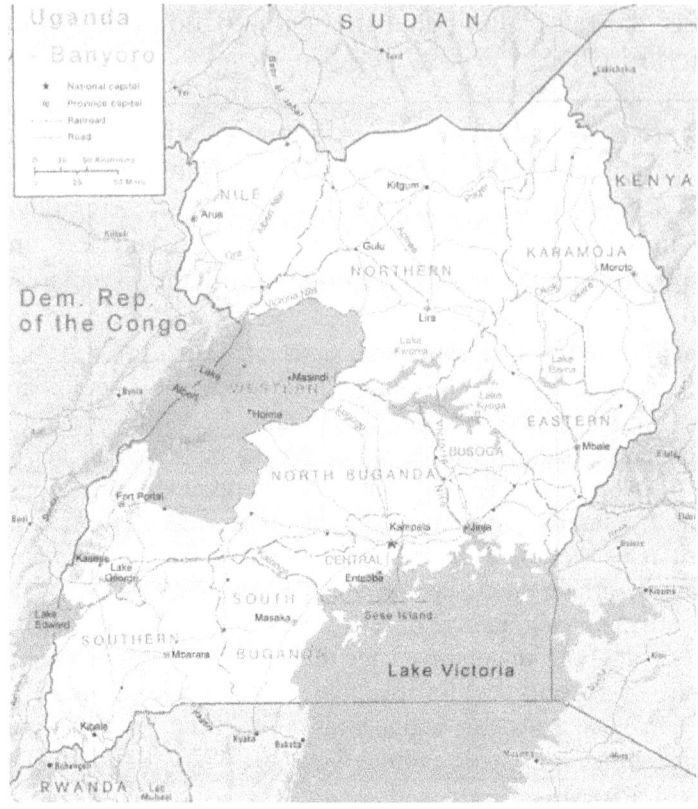

Bunyoro is a kingdom in Western Uganda. It was one of the most powerful kingdoms in East Africa from the 16th to the 19th century. It rose to power by controlling a number of the holiest shrines in the region, the lucrative Kibiro saltworks of Lake Albert, and having the highest quality of metallurgy in the region. The Kingdom of Bunyoro-Kitara is the remainder of a once powerful empire of Kitara. At the hight of its glory, the empire included present day Masindi, Hoima, Kibaale, Kabarole and Kasese districts; also parts of present day Western Kenya, Northern Tanzania and Eastern Congo.

Ethiopian Empire (1137–1974)

The Ethiopian Empire also known as Abyssinia, covered a geographical area that the present-day northern half of Ethiopia covers. It existed from approximately 1137 (beginning of Zagwe Dynasty) until

1975 when the monarchy was overthrown in a coup d'état. In 1270, the Zagwe dynasty was overthrown by a king claiming lineage from the Aksumite emperors and, hence, Solomon. The thus-named Solomonic Dynasty was founded and ruled by the Habesha, from whom Abyssinia gets its name.

The Habesha reigned with only a few interruptions from 1270 until the late 20th century. It was under this dynasty that most of Ethiopia's modern history occurred. During this time, the empire conquered and incorporated virtually all the peoples within modern Ethiopia. They successfully fought off Italian, Arab and Turkish armies and made fruitful contacts with some European powers, especially the Portuguese, with whom they allied in battle against the latter two invaders.

(from Atlanta Blackstar https://atlantablackstar.com/2013/12/05/7-midieval-african-kingdoms/3/

Southern Africa

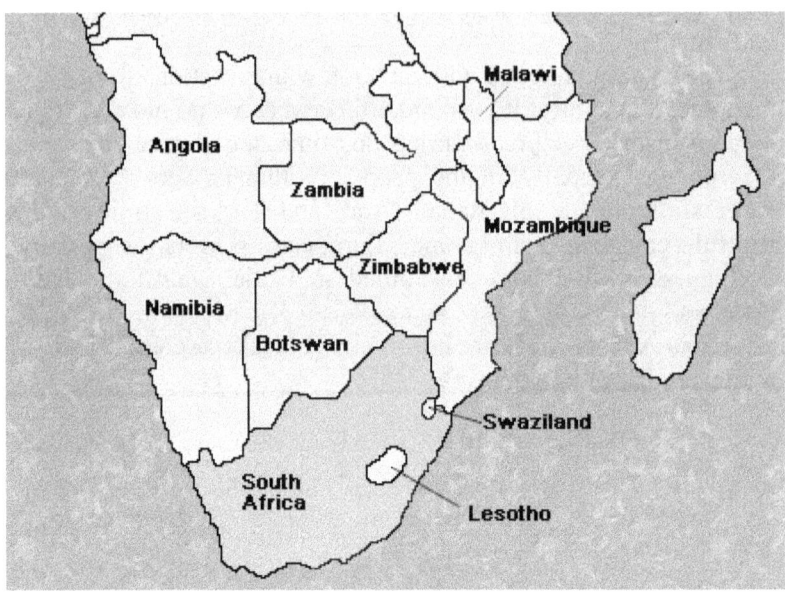

The **Kingdom of Mapungubwe (or Maphungubgwe)** (c.1075–1220) was a state in South Africa located at the confluence of the Shashe and Limpopo rivers, south of **Great** Zimbabwe. The name is derived from either TjiKalanga and Tshivenda. The name might mean "Hill of Jackals".[1] The kingdom was the first stage in a development that would culminate in the creation of the Kingdom of Zimbabwe in the 13th century, and with gold trading links. (Wikipedia)

They traded gold and ivory with China and India and had a flourishing agricultural industry.

Mapungubwe is also the earliest known site in southern Africa where evidence of a class-based society existed. This means that the leaders were separated from the commoners.

Findings in the area are typical of the Iron Age and smiths created objects from iron, copper and gold for local use, as well as for trading. Pottery, wood, ivory, bone and shells indicate that many other materials were used and traded with other cultures. The most spectacular of all the discoveries found at the site is a small, golden rhinoceros, made of gold foil with a wooden core.

Mapungubwe's prosperity faded in the 1300s due to climate change, which in turn led to migrations further north, towards Zimbabwe.

The **Mutapa Empire or Empire of Great Zimbabwe** (1450–1629) was a kingdom located between the Zambezi and Limpopo rivers of Southern Africa in the modern states of Zimbabwe and Mozambique. Remnants of the historical capital are found in the ruins of Great Zimbabwe. The **Kingdom** of **Zimbabwe** (c. 1000–1450) was a medieval Shona (Karanga) **kingdom located** in modern-day **Zimbabwe**. Its capital, Lusvingo, now called **Great Zimbabwe**, is the largest stone structure in precolonial Southern Africa. (from *Ancient History Encyclopedia*)

The first inhabitants of Great Zimbabwe were Shona-speaking peoples who likely settled in the region as early as 400 C.E. Back then, the land was full of possibilities: plains of fertile soil to support farming and herding, and mineral rich territories to provide gold, iron, copper, and tin for trading and crafting. It was fine place for the Shona to call home.

Discoveries of Chinese porcelain, engraved glass from the Middle East, and metal ornaments from West Africa provide evidence that Great Zimbabwe participated in a comprehensive trade network during the 13th and 14th centuries. Gold was probably its chief export and East African cities — especially those along the coast that had overseas connections — were most likely its primary trading partners.

The wealth of Great Zimbabwe lay in cattle production and gold. There are a number of mines to the west of Great Zimbabwe, about 40 kilometres away. One theory is that the rulers of Great Zimbabwe did not have direct control over the gold mines, but rather managed the trade in it, buying up huge quantities in exchange for cattle.

Moors Influence in Europe

Golden Age of the Moor, edited by Ivan Van Serima

Page 9.....

2. Influences and Contributors

A distinction should be drawn between the Classical Renaissance of Europe, which maily relates to its literature and art, and the Scientific Renaissance, which began to bud and flower in the 12th and 13th centuries. Jose Pimienta-Bey deals primarily with the Moorish stimulus for the latter. He sets out to prove in his essay, and does so with a formidable body of evidence, that the foundation of much of medieval Western science and its academies was built up upon the transmissions, refinements and and discoveries of the Arabs and the Moors.

Moorish influence came primarily to the West by the way of the Iberian peninsula (al-Andalus by the Moors).

Pg 10....

Nealy all the major universities in Europe sprung up around the same time, beginning in the second half of the 12th century right up through the 13th, a span of about one hundred and fifty years, a period which coincides with the flowering of Moorish science and the establishment of centers in Europe to translate Moorish treatises from Arabic to Latin. In Italy, we have Bologna, Padua, Naples, Rome; in France, Montpelier and Toulouse; in Portugal, Lisbon and Coimbra; in England, Oxford.

Pg 13....

"At a time (around 1492) when the most insignifican provinces of Moorish Spain contained libraries running into thousands of volumes, the cathedrals, monasteries and palaces of Leon, under Christian rule, numbered books only by the dozen. The paltry number of texts the Christians did possess were almost all devotional of liturgical."

The narrowness of vision this produced among leaders of the church and state was to have catastrophic effects. It led to the massive burning of African and books under of Cardinal Ximenes de Cisneros. It inspired a similar bonfire of the books of native Americans. Bishop de Landa exhorted his followers in the Yucatan "Burn them all – they are works of the devil."

.....Hatred of the dark invaders kindled the bonfires. The Church at that time too saw most of this foreign learning as something evil. Even demonic......

Pg 28.....

During the Middle ages, because of his dark complexion and Islaamic faith, the Moor became in Europe a symbol of guile, evil and hate. In medieval literature demonic figures were commonly depicted with black faces. Among Satan's titles in medieval folklore were: "Black Knight," "Black Man," "Black Ethiopian," and Big Negro." In the Cantiga 185 of King Alfonso the Wise of Spain (1254 – 86), three Moors attacking the Castle of Chincoya are described as "black as Satan."

Pg 376.....

Though the Moors lost militarily, political and economic control of Spain, their influence lingered long after their physical departure.

Pg 382,,,,,,

CAIRO: SCIENCE ACADEMY OF THE MIDDLE AGES (by Beatrice Lumpkin and Siham Zitzler)

Summary: In the Middle ages, Egypt and North Africa continued their tradition of leadership in science and mathematics, a tradition then already 4,000 years old. At Cairo, a Science Academy was established, similar to the Science Academy of Baghdad. From North Africa, the most advanced mathematics, science medicine and literature were introduced to less developed Europe.......

****

Technology

The Muslims 'Moors' were able to pass on knowledge they acquired from China, India, and other regions of the world they had come across through trade. New technology was introduced, such as the Arabic numeral system, paper from China, the astrolabe, and their irrigation system, which aided in growing crops in Spain that had been unable to grow before. Their farming methods and irrigation system, which were established due to living in the deserts of North Africa, were able to help Spain's agriculture flourish. The Moors were also able to introduce new crops such as citrus and sugarcane into the region. Since then, Spain has become one of the largest producers of citrus fruit.

(This article was written by Mena Mahmoud)

****

Beginning in the Renaissance, "**Moor**" and "blackamoor" were also used to describe any person with dark skin. In A.D. 711, a group of North African Muslims led by the Berber general, Tariq ibn-Ziyad, captured the Iberian Peninsula (modern Spain and Portugal)

****

From Black History Studies (https://blackhistorystudies.com/resources/resources/15-facts-on-the-moors-in-spain/)

15 Things You Did Not Know About the Moors of Spain

1. The Spanish occupation by the Moors began in 711 AD when an African army, under their leader Tariq ibn-Ziyad, crossed the Strait of Gibraltar from northern Africa and invaded the Iberian peninsula 'Andalus' (Spain under the Visigoths).

2. A European scholar sympathetic to the Spaniards remembered the conquest in this way:

a. [T]he reins of their (Moors) horses were as fire, their faces black as pitch, their eyes shone like burning candles, their horses were swift as leopards and the riders fiercer than a wolf in a sheepfold at night . . . The noble Goths [the German rulers of Spain to whom Roderick belonged] were broken in an hour, quicker than tongue can tell. Oh luckless Spain! [i]

[i] Quoted in Edward Scobie, The Moors and Portugal's Global Expansion, in Golden Age of the Moor, ed Ivan Van Sertima, US, Transaction Publishers, 1992, p.336

3. The Moors, who ruled Spain for 800 years, introduced new scientific techniques to Europe, such as an astrolabe, a device for measuring the position of the stars and planets. Scientific progress in Astronomy, Chemistry, Physics, Mathematics, Geography and Philosophy flourished in Moorish Spain

4. Basil Davidson, one of the most noted historians recognized and declared that there were no lands at that time (the eighth century)

"more admired by its neighbours, or more comfortable to live in, than a rich African civilization which took shape in Spain"

5. At its height, Córdova, the heart of Moorish territory in Spain, was the most modern city in Europe. The streets were well-paved, with raised sidewalks for pedestrians. During the night, ten miles of streets were well illuminated by lamps. (This was hundreds of years before there was a paved street in Paris or a street lamp in London.) Cordova had 900 public baths – we are told that a poor Moor would go without bread rather than soap!

6. The Great Mosque of Córdoba (La Mezquita) is still one of the architectural wonders of the world in spite of later Spanish disfigurements. Its low scarlet and gold roof, supported by 1,000 columns of marble, jasper and and porphyry, was lit by thousands of brass and silver lamps which burned perfumed oil.

7. Education was universal in Moorish Spain, available to all, while in Christian Europe ninety-nine percent of the population were illiterate, and even kings could neither read nor write. At that time, Europe had only two universities, the Moors had seventeen great universities! These were located in Almeria, Cordova, Granada, Juen, Malaga, Seville, and Toledo.

8. In the tenth and eleventh centuries, public libraries in Europe were non-existent, while Moorish Spain could boast of more than seventy, of which the one in Cordova housed six hundred thousand manuscripts.

9. Over 4,000 Arabic words and Arabic-derived phrases have been absorbed into the Spanish language. Words beginning with "al," for example, are derived from Arabic. Arabic words such as algebra, alcohol, chemistry, nadir, alkaline, and cipher entered the language. Even words such as checkmate, influenza, typhoon, orange, and cable can be traced back to Arabic origins.

10. The most significant Moorish musician was known as Ziryab (the Blackbird) who arrived in Spain in 822. The Moors introduced earliest versions of several instruments, including the Lute or el oud, the guitar or kithara and the Lyre. Ziryab changed the style of eating by breaking meals into separate courses beginning with soup and ending with desserts.

11. The Moors introduced paper to Europe and Arabic numerals, which replaced the clumsy Roman system.

12. The Moors introduced many new crops including the orange, lemon, peach, apricot, fig, sugar cane, dates, ginger and pomegranate as well as saffron, sugar cane, cotton, silk and rice which remain some of Spain's main products today.

13. The Moorish rulers lived in sumptuous palaces, while the monarchs of Germany, France, and England dwelt in big barns, with no windows and no chimneys, and with only a hole in the roof for the exit of smoke. One such Moorish palace 'Alhambra' (literally "the red one") in Granada is one of Spain's architectural masterpieces. Alhambra was the seat of Muslim rulers from the 13th century to the end of the 15th century. The Alhambra is a UNESCO World Heritage Site

14. It was through Africa that the new knowledge of China, India, and Arabia reached Europe. The Moors brought the Compass from China into Europe.

15. The Moors ruled and occupied Lisbon (named "Lashbuna" by the Moors) and the rest of the country until well into the twelfth century. They were finally defeated and driven out by the forces of King Alfonso Henriques. The scene of this battle was the Castelo de Sao Jorge or the 'Castle of St. George.'

****

From Wikipedia

(htttps://en.wikipedia.org/wiki/List_of_inventions_in_the_medieval_Islamic_world)

List of inventions made in the medieval Islamic world

There is a list of inventions made in the Islamic world, especially during the **Islamic Golden Age**, as well as in later states of the Age of the Islamic Gunpowders such as the Ottoman and Mughal empires.

The **Islamic Golden Age** was a period of cultural, economic and scientific flourishing in the history of Islam, traditionally dated from the eighth century to the fourteenth century, with several contemporary scholars[who?] dating the end of the era to the fifteenth or sixteenth century. This period is traditionally understood to have begun during the reign of the Abbasid caliph Harun al-Rashid (786 to 809) with the inauguration of the House of Wisdom in Baghdad, where scholars from various parts of the world with different cultural backgrounds were mandated to gather and translate all of the world's classical knowledge into the Arabic language and subsequently development in various fields of sciences began. Science and technology in the Islamic world adopted and preserved knowledge and technologies from contemporary and earlier civilizations, including Persia, Egypt, India, China, and Greco-Roman antiquity, while making numerous improvements, innovations and inventions.

There is a List of inventions, which is incomplete

This list expands from the 7th – 17th centuries.

(1700 AD – 1875 AD)

Period of Colonializers' Disruptions

Africa Diaspora Beginnings

In 1964, UNESCO launched the elaboration of the General History of Africa (GHA) with a view to remedy the general ignorance on Africa's history. The Collection: Vol. I – XI
https://en.unesco.org/general-history-africa

****

U N E S C O General History of Africa, Volume V: Africa from the Sixteenth to the Eighteenth Century (Editor B . A . Ogot)

In 1500 the geo-political m a p of the world revealed the existence of a number of major, relatively autonomous regions which were to some degree interlinked either through trade or through conflict. First, there was the Far East represented by Japan and China which, together with the Pacific and Indian Ocean regions covering the Molucca Islands, Borneo, Sumatra and India itself, were the world's source of spices. Next, was the Middle East which covered a vast area including the Arabian peninsula, the Safawid empire and the Ottoman empire which was soon to include North Africa. Then there was Europe with its Slavs, Scandinavians, Germans, Anglo- Saxons and Latins, all of w h o m still remained confined within its borders. Finally, there was Africa with its Mediterranean seaboard in the north and its Red Sea and Indian Ocean coastlines which were becoming increasingly involved in the international trade with the Far East and the Orient.

The period from 1500 to 1800 was to witness the establishment of a new Atlantic-oriented geo-economic system, with its triangular trading pattern linking Europe, Africa and the Americas. With the opening up of Atlantic trade, Europe - particularly Western Europe — gained ascendancy over the Americas and African societies. Henceforth, Europe was to play a leading role in the accumulation of capital generated by trade and plunder on a worldwide scale. The emigration of Europeans to trading settlements in Africa and the territories of North and South America gave rise to the establishment of supporting overseas economies. These were to play a decisive long-term role through their contribution to Western Europe's rise to power over the rest of the world.

Throughout the 1400s and early 1500s, the main market for 'the black merchandise' was Europe, particularly Portugal and the Spanish countries and, to a certain extent, islands in the Atlantic such as Madeira, the Canaries, the Cape Verde Islands and subsequently St Thomas Island — although the number of slaves transported to these islands was limited by the small size of the islands themselves. The main incentive for the slave trade in Madeira, the Cape Verde Islands and, in particular, St Thomas Island was the introduction of the cultivation of sugar cane and cotton. Slavery could not develop to any great extent on the European continent because there was no economic reason for it. The Africans who were brought into Portugal and the Spanish countries were mainly employed as domestic servants or semi-skilled artisans in the towns.

From the point of view of world history, the export slave trade from Africa, particularly the trans-Atlantic trade, is unique in several respects. The sheer size of the trade, the geographical extent of the regions of the world involved, and the economics of the trade - at the level of slave supply, employment of slaves to produce commodities for an international market, and trade centred on the products of slave labour - all these put the slave trade from Africa apart from all other slave trades.

The British North American colonies moved gradually from subsistence activities to market production under the impact of the Atlantic system, three economic regimes comprising the southern, middle and northern (mainly New England) colonies became distinguishable. The combination of natural resources and the availability of cheap African slave labour encouraged the southern colonies to expand plantation agriculture, first rice and tobacco, then cotton. The middle colonies, however, took to foodstuffs production on family-sized farms utilizing family labour. In the northern colonies, relatively poor natural resources for agriculture combined with the availability of deep natural harbours and forest resources for shipbuilding encouraged early specialization in trade and shipping.34 In this way, the south produced virtually all the plantation commodities that were exported to Europe, while the north produced the bulk of the invisible exports - shipping, merchants' services, insurance, etc. The middle colonies, meanwhile, produced footstuffs for export as well as some invisible exports. Production in the south was dependent on

African slave labour, while the market for the output was mainly in Europe. For the middle and northern colonies, the restructuring of the Caribbean economies that accompanied the growth of slave plantations produced a division of labour between the Caribbean (British and non-British) and North America which provided a large market in the Caribbean for the foodstuffs of the middle colonies and for the shipping and other services of the northern colonies. Thus, the economies of the three North American sub-regions were tied up with the slavery system

The virtual elimination of the entire Indian population led to two important results. The first is that the phenomenal expansion of commodity production for sea-borne trade with Europe and North America – which occurred between the sixteenth and nineteenth centuries - was made possible only by a massive import of African slave labour. The second is that the agricultural land in Latin America and the Caribbean was taken over by European colonists and put into large estates that came to be known as *ladende* or *lazenda.* It will be shown subsequently that both developments provided trading opportunities that stimulated the capitalist transformation of Western Europe and North America, while at the same time producing underdevelopment and dependence in Latin America and the Caribbean.

It was this massive transplantation of African labour into Latin America, the Caribbean Islands and the southern territories of North America that produced the phenomenal expansion of commodity production and trade in the Atlantic area between the sixteenth and nineteenth centuries. This, in turn, provided the stimulating opportunities and challenges under pressure of which the process of capitalist transformation was completed in the major West European countries and North America. This same historical process, however, produced structures of underdevelopment and dependence in Latin America and the Caribbean.

The establishment of a seaborne commerce between Africa and Western Europe from the second half of the fifteenth century seemed at first to offer the kind of opportunities that black Africa needed for rapid economic and social transformation. The gold trade expanded. Trade in agricultural commodities, such as pepper, was initiated. Some stimulus was even given to African cloth producers as the Portuguese and the Dutch participated in the distribution of African cloths to different parts of the African coast.61

These early developments, however, were short-lived. When the vast resources of the Americas became accessible to Western Europe after 1492, and with the virtual elimination of the Indian population as a consequence of the conquest and the diseases introduced by the European conquerors, the role of Africa in the evolving Atlantic economic system was altered. The population which Africa needed to build up in order to provide the internal conditions necessary for a complete structural transformation of the continent's economies and societies was transferred massively to the Americas where it was employed to develop large-scale commodity production and trade. The conditions created over a period of about three centuries by this massive transfer of population discouraged the development of commodity production in Africa, both for internal trade and for export, and laid the foundation for dependency structures in the continent.

U N E S C O General History of Africa, Volume VI: Africa in the Nineteenth Century until the 1880s (Editor J. F . A . Ajayi)

The great transformation of Africa's economic relations with the wider world did not occur with the late nineteenth-century partition by European powers. Rather, conversely, the partition of Africa was a consequence of the transformation of Africa's economic relations with the wider world, and in particular with Europe, a transformation that took place in a period beginning approximately in 1750 and culminated in the extensive European direct colonization of the last decades of the nineteenth century.

There had long been trading networks in various parts of Africa, and many of these networks had extended beyond the frontiers of the African continent - across the Indian Ocean, the Mediterranean, and the Atlantic. By and large, these extra-continental trading links constituted the same kind of iong-distance trade' as that which had been well known for millennia in Asia and Europe as well. Such long-distance trade involved the exchange of so-called luxury products, that is products that were usually small in bulk and high in profit per unit of size. The production of such items for exchange tended to involve a small proportion of the manpower of the originating zones and probably a small proportion of the total value produced in these zones. In these senses, 'luxury' trade was 'non-essential' trade in that its interruption or cessation did not require any basic reorganization of productive processes in the originating zones. T h e two zones whose products were thus exchanged could not therefore be said to have been located in a single social division of labour.

****

http://www.unesco.org/new/fileadmin/MULTIMEDIA/HQ/CLT/pdf/P_Lovejoy_African_Contributions_Eng_01.pdf
Collective Volume The Slave Route Project, UNESCO

African Contributions to Science, Technology and Development by Paul E. Lovejoy

Introduction

Scientific discovery and the application of technology to the natural environment have been essential to the history of Africa and in the development of the African Diaspora throughout the world, and especially in the Americas.

When Africans migrated, whether under conditions of slavery or as voluntary travelers, they took with them knowledge of agricultural techniques and skills in exploiting the nature environment that were necessary for development.

The African contribution to science and technology can be appreciated with respect to the impact on the development of the Americas, which suffered severe population destruction through disease and European conquest after 1492. Spain, Portugal and then other western European countries took advantage of military superiority and the demographic catastrophe in the Americas to confiscate vast tracts of land, which only needed labor and transferred technology for its development.

Europeans empires and the generation of enormous wealth depended upon the combination of these ingredients – virtually free and very fertile land, labor and technology, largely from Africa, and the ability to garner huge profits through the reliance on slavery, in which workers were not paid for their labor or their technology.

It is crucial to note that none of the major plantation crops in the Americas and only a few of the foodstuffs consumed by people in the Americas came from western Europe, while virtually all of the newly introduced crops originally came from Africa or were grown there before their introduction to the Americas.

Sugar cane was first grown in the Mediterranean and in southern Morocco before spreading to other offshore islands and then to the Americas.

Cotton was grown and made into textiles in the western Sudan and in the interior of the Bight of Benin for centuries before being introduced to the Americas, along with weaving, indigo dyeing, and the decorative arts associated with textiles.

Rice, indigenous to West Africa, was introduced into the sea islands of South Carolina and Georgia, as well as the Mississippi valley, Maranhão in northeastern Brazil, and elsewhere, while numerous foodstuffs and stimulants were transferred from Africa as well.

As Judith A. Carney and Richard Nicholas Rosomoff have established that Africans established "botanical gardens of the dispossessed," in which they cultivated many familiar foods, including millet, sorghum, coffee, okra, watermelon, and the "Asian" long bean, for example, all of which were native to Africa.

We now know that indigenous knowledge of botany and zoology was crucial in the evolution of modern science.

Similarly, the application of technology to develop such commercial products as Coca Cola, Worcestershire Sauce, and palm-oil based soap rely on African plants that first came to the notice of Europe and the Americas through the activities of slave ships.

While it may have been possible that these crops and skills would have reached the Americas anyway, the fact is that the transfer of technology and the accumulation of scientific knowledge underlying that technology were done under conditions of slavery.

There are many problems with the stereotypes of African under-achievement that have to be addressed. First, it is not true that Africa was a continent of little achievement. We can re-examine the contributions in various areas, especially agriculture, mining, and pharmacology, that will demonstrate that the contributions in these areas alone have been important. Moreover, it is argued here, we must overcome narrow conceptions of science and technology that undermine our understanding of development – how things happen and what it has taken in terms of human intellect to achieve practical goals and improve standards of life.

A review of African contributions to science, technology and development asks the crucial question: why was slavery necessary in the "development" of the Americas? Wasn't it possible to have achieved similar or even better results without slavery? Why has it been popular to conceive of development as a product of science and technology, as if slavery has not existed? Would not the scientific and technological development of the modern world have been greater without slavery? These questions frame this discussion of African contributions to development and challenge the stereotype that Africans have made only marginal, if any, contributions to modern science and technology. Rather, it can be shown that slavery stripped Africa of more than just people who might have helped in the development of Africa, but the forced migration also confiscated technological knowledge that was based on previous experience and training that was transformed in the Americas.

Slavery was particularly inefficient in allocating skilled people to where their skills were needed most, but this was the way technology was transferred for purposes of development to benefit the few.

The failure to recognize African contributions to science and technology demonstrates that conceptions of scientific knowledge have been racialized, as if knowledge and discovery bear some correlation with the color of skin. The classification of nature and the exploration of the environment for practical application are universal, whether in the Arctic or the tropics and whether in Africa, Asia or Europe. The development of the modern disciplines of botany, zoology, pharmacology, and medicine has occurred because of classification, comparative methodologies that emphasize the discipline of observation, and experimentation that attempts to verify results. To some extent what is often considered to be modern science has developed in isolation from bodies of knowledge that were developed in non-western locations, such as China, the Islamic world, the indigenous Amerindian populations of the Americas, and Africa.

Modern science to some extent has been based on a racialized premise: if it is not "discovered" in a European or North American laboratory, there has been no discovery and it is not science. It is only recently, that "scientists" have increasingly turned to bodies of knowledge that derive from alternate systems of classification and analysis.

A botanical vocabulary in Yoruba, based on scientific knowledge collected in Bahia by Pierre Verger, not in Nigeria, is over 700 pages. If the knowledge of botany in Nigeria were added to this compendium, the encyclopedia would be more extensive. This scientific knowledge has considerable pharmacological significance. Similarly, the chemical composition of the numerous salts of the central Sahara and the Lake Chad basin was generally understood in terms of the applications in pharmacology, cuisine, tanning, textile dyeing, and veterinary care. The distinctions between sulfates, carbonates, and chlorides was recognized, and to some extent efforts were made to isolate these various salts in a manner that reveals a level of scientific enquiry that was certainly transferrable to the Americas. Both Yoruba botany and the scientific knowledge underlying Central Sudan salts demonstrate a scientific sophistication that was transferred within West Africa and to the Americas and the African Diaspora there.

The Dawn of Civilization

African contributions to the ancient world are well known. The pyramids of Egypt and Kush attest to the skills of engineering and architecture. Classical Egypt crossed all the frontiers of northeastern Africa and southwestern Asia. The populations of this area were mixed. Those who built the pyramids included Africans from the middle and upper Nile River valley, as well as people from the Mediterranean and elsewhere. The technology and the science behind the technology were not racialized but crossed many cultures. Similarly, the Nok culture of what is now central Nigeria displays an antiquity in art forms that reveal knowledge of metallurgy and stone sculpture that has similarities to other parts of the world. This is important to recognize; technological and scientific breakthroughs occurred independently in many parts of the world.

The spread of iron technology is a case in point. Africans could transfer the skills of blacksmithing to the Americas because these skills were ancient in Africa. If anything, some African skills were not transferred into diaspora, such as the ability to work in other metals, including bronze and silver. Generally, there was no need for these skills in plantation America, while there was a need for skills in working iron. This is but one example of the types of technological knowledge that were common in Africa but were not transferred into diaspora but rather the skills were lost, retarding development. Once again we see the inefficiency of slavery as a system; the exploitation of people as slaves tended to undermine the transfer of skills.

The construction of ancient monuments, palaces and temples in the Nile valley demonstrates an architectural tradition that was continued in the construction of mosques in West Africa and along the East African coast, as well as churches in Ethiopia. **The knowledge of mathematics and engineering** is ancient and was closely tied to the availability of building materials. In the Americas, architectural contributions can be seen in the construction of forts and churches, especially in Cuba and mainland Latin America, where Africans and people of African descent were involved in both construction and maintenance. Many of the palaces, mosques, temples, churches, and fortifications have been designated UNESCO Heritage sites from medieval times onward. The ancient pyramids of the Nile valley demonstrate that the issue of what was "African" and what was not is a question of definition. Certainly the pyramids are unique and as much a part of the history of African contributions to technology and development.

The Domestication of Agriculture

It is clear that African contributions to agriculture have underpinned the development of the African continent and to a great extent the Americas as well. This can be seen with respect to several agricultural innovations, including the domestication of millet/sorghum, rice, yams, kola, and coffee. The base line for understanding the contribution of Africans to agricultural development, both in Africa and then in the Americas, is the emergence of civilizations that are reflected in the sculpture of the Nok complex in Nigeria and the enclosures of Zimbabwe in southern Africa. Agriculture, including animal husbandry, evolved independently in Africa, which in a real sense was not only the origin of all people but also the cradle of food production, crop specialization, and experimentation in systems of agriculture and transhumance livestock management.

In the Americas, the cultivation of root crops in plantation settings was often done on the side, not as part of the work load for the slave owners but for the subsistence strategies of the African population. In time of food shortage or natural disasters, such as hurricanes, survival depended upon access to root crops. Agricultural innovation does not just involve the development of new crops but also the adoption and adaptation of imported species of foodstuffs, including bananas, maize, and manioc, and their significance in terms of agricultural production and sustainability. Similarly, farmers inevitably experimented with different strategies of agricultural cultivation, from swidden systems to irrigation, tree and root crops, grains, fruit, vegetables, condiments and spices.

The Bambara, Fula, Malinke, and Songhay had long experience in growing rice along the Niger River, while Serer, Mende, Temne, Kissi, Papel, and Baga utilized their own special techniques of rice production from Senegal to the Ivory Coast. There are extensive scholarly studies of rice-producing societies on the upper Guinea coast, including the Doala, Bran, and others.

One of the chief contributions of enslaved Africans to the Americas was the transmission of the West African rice knowledge systems – land-use principles, gendered division of labor, and processing techniques.

Many enslaved Africans were experienced rice producers, and as a result British colonists in South Carolina had some difficulty in cultivating the crop successfully. Many of the practices of early production in South Carolina paralleled those in Africa. Thus enslaved Africans were active rather than passive participants in the founding of American civilization.

Africans played an important role in the development of the commercial rice industry in colonial South Carolina and Georgia. Enslaved laborers on South Carolina rice plantations were skilled. Throughout the eighteenth century, planters placed a positive value on slaves brought from rice-growing regions, which is revealed in newspaper advertisements by South Carolina planters searching for runaway slaves. Colonists preferred captives from the Rice Coast, despite the predominance of Congo-Angola slaves. Merchants in Charleston and Savannah maintained a close relationship with the owners of the Bance Island in the Sierra Leone River, whose captives often were destined for rice plantations in South Carolina and Georgia. Planters using enslaved Africans subsequently transmitted rice-growing technology to Texas, Louisiana, and Brazil.

Stimulants

Africa was the origin of three important pharmacological substances, known as alkaloids (kola, coffee, khat).

The principal characteristic of these agricultural commodities is that they have pharmacological properties that stimulate the brain and the central nervous system. Like other alkaloids (tea, cacao, betel, coca, tobacco, opium, etc.), they have virtually no food value but rather provide the sensation of reducing hunger and fatigue. The active ingredients in the various alkaloids vary, caffeine (kola, coffee) and theobromine (tea, kola) being two of the most important.

Kola nut production, particularly Cola nitida, is indigenous to western Africa and is the basis of the popular cola drinks, which were developed in the United States and Europe in the 1880s and 1890s. .

Kola nuts, which are eaten because they contain caffeine, theobromine and kolatin, are a popular stimulant in many parts of West Africa. Like other mild stimulants, including coffee, tea and cocoa, kola nuts are moderately addictive.

Kola is indigenous to the West African forest, but is found as far east as Gabon and the Congo River basin. Of its more than forty varieties, four – C.nitida, C. acuminata, C. verticillata, and C.anomala – are the most common of the edible species, and have been important in the commerce of West Africa. These four types are similar in their chemical composition and use. They contain, together with other compounds, large amounts of caffeine, and smaller quantities of theobromine, kolatin, and glucose. All these are stimulants: caffeine affects the central nervous system, theobromine activates the skeletal muscles, kolatin acts on the heart, and glucose provides energy to the body as a whole.

Coffee was developed as a consumable drink in Ethiopia and subsequently spread to Arabia and from there throughout the world. Coffee contains several compounds, particularly caffeine, but also theophylline and theobromine, each of which has different biochemical effects on the human body. The major compound in coffee is caffeine, which is a mild stimulant that can enhance alertness, concentration and mental and physical performance.

Manufacturing and Industry

Since at least 1,000 CE, and in some places there was also local production of raffia cloth and the use of skins and leather as clothing. Most significant development, however, was the manufacture of cotton textiles. An examination of the textile industry in West Africa demonstrates indigenous innovation and development. This can be seen clearly in the development of indigo dyeing, including the development of dye pits using locally produced cement. It should be noted that the production of indigo for industrial purposes spread to the Americas via the African diaspora. Similarly, cotton production was also developed in the Americas, and especially in the United States in the nineteenth century.

Cotton had been cultivated, harvested, cleaned, spun into yarn, and woven into cloth for centuries in West Africa, long before direct trade with Europeans. Cotton was being grown in the region of Senegambia and probably across the whole of the savanna by 1000 A.D. Consumer markets for certain types of textiles were already in place when Portuguese mariners began exploring the coastline in the fifteenth century. Early centers of textile production arose in at least two areas of West African Interior. A western centre was located around the upper Niger, Gambia and Senegal watersheds, and contiguous areas on the desert edge. An eastern centre was located around Lake Chad, and the area of the early Hausa kingdoms.

Weaving is a craft based on calculation and counting. In order to set up a loom, the weaver must estimate how much thread is needed for the warp, which depends on the type of thread, its thickness or fineness, and the intended dimensions and density of the fabric. Then the warp threads have to be measured out to the correct length, placed in sequential order, and arranged on the loom. How the warps are arranged and manipulated creates different types of fabric structures.

Indigo has played an important role in local, regional, and international economic histories. Indigo is native to native tropical areas of south Asia, the Americas and Africa. Various parts of the plant yield medicines and dyestuff that came to be known and highly valued in the ancient Mediterranean world and were imported by the Greeks and Romans by a least the last few centuries BC. Indigofera tinctoria, was native to eastern and southern Africa, though it has been widely cultivated in West Africa as well, and it was vigorously promoted in Nigeria after 1905 as a richer source of dye than I. tinctoria

Production and trade are, of course, a two way street. By weaving, dyeing, sewing and embellishment such a rich variety of cloth and clothing, African artisans helped to facilitate the workings of local regional and international commerce over time. And along the way, they were able to gain access to new fibers, techniques and imagery which opened up possibilities for the invention of new products and the alteration of standard ones. In other words, the phrase "local craft production" is something of an oxymoron, for textile manufactures were neither insulated from external change nor impervious to it. One factor that contributed much to this dynamism was an especially high cultural value placed on cloth. Peoples in West Africa, and in many others parts of the continent, shared an avid appreciation of the powerful sensual qualities of woven fabric and the serious matters of proper and stylish dress and public display.

Salt Production and Pharmacology

An examination of salt production and pharmacology in the central Sudan of West Africa demonstrates the impact of technological developments in West Africa. As the chemistry of various salts that were exploited suggests, there was considerable knowledge of NaCl (sodium chloride), sodium sulfates, potassium chlorides and sulfates. Different types of salt, including natron, trona, vegetable salts, and sea salt, had significance for use as medicines, culinary purposes, tanning of leather, and other uses.

Salt was scarce in Africa before the twentieth century.

Salt was found in scattered deposits, mostly in the Sahara and in the desert area near the Red Sea but also released through brine springs in widely scattered locations. Salt was also extracted from sea water through evaporation.

The most sophisticated production of salt developed in the Central Sudan, and particularly in the area dominated by the state of Borno, and its predecessor, Kanem, in the basin of Lake Chad.

The medicinal uses of the different salts were numerous: ungurnu, or trona, from the eastern shores of Lake Chad, white natron from Mangari and Kawar, and red natron from Mangari and Kawar contained high concentrations of sodium carbonates and hence were excellent for stomach ailments. Local medicinal knowledge credited the different types of natron with specific properties: some were milder and better for children and elders, while others were useful in pregnancy. Because Mangari salt was so similar to natron, it, too, could be used as medicine. In addition, natron and varieties of Mangari salt were used in various mixtures to treat dandruff, problems related to pregnancy, eye disorders, infertility, and as an ingredient in curative potions and mixtures.

Culinary uses were equally specialized. Specific recipes required their own salt or natron. The standard Hausa millet porridge, for example, could be made with various grades of white natron, ungurnu, red natron, or Mangari salt; each recipe had a different name.

Salt and natron were also mixed with tobacco, which was commonly chewed or taken as snuff. Salt, whether from Kawar, Mangari, or Lake Chad, was used widely for this purpose in the Central

Sudan, Asante, the Yoruba states, and elsewhere. Ungurnu from Lake Chad was especially popular in Asante, but white natron from Muniyo was also common. Any salt could be added to bring out a pungent taste to the tobacco, and preference appears to have varied with the consumer and availability.

Conclusion: African Contributions to Science and Technology

This exploration of technological innovation demonstrates the dimensions of African contributions to scientific discovery. A full historical analysis must await further research. The extensive writings of West African scholars at Timbuktu and other places in the Sahel and savanna undoubtedly contain important information of past discoveries, especially in mathematics. Moreover, the achievements of the Dogon in astronomy and the contributions to navigation in the Indian Ocean and Red Sea, and Atlantic are further examples, just as the knowledge of the chemistry of salt, iron, copper, tin and gold were well understood in Africa, long before any direct trade with Europe.

The failure to recognize African contributions to science and technology and the transfer of expertise to the Americas minimizes the role of enslaved Africans in the development of the Americas, despite the often inefficient use of the skills and talents of individuals. Africans and their descendants were primarily responsible for the production of staple export crops and the mining of gold and silver. They also had to feed themselves, which to a great extent relied on food crops and recipes brought from Africa. African labor was important in maritime commerce and in the activities of port towns throughout the Atlantic world. Even where technology was known in Europe, such as blacksmithing, many smiths in the Americas were from Africa, where they brought similar technological expertise and skill in metallurgy. European technology excelled in weaponry and naval wares. Otherwise, whether in textile and leather production, agriculture, and mining, there was little technological contribution from Europe that superseded that brought from Africa, at least before the nineteenth century.

Was it necessary, therefore, to secure the development of the Americas through the confiscation of land, the appropriation of labor and technology through slavery, and the use of military and naval superiority to subjugate people? The violent concentration of wealth through the gains from the exploitation of tropical production was based on enslaved labor, mostly from Africa. The gains were instrumental in the emergence of banking, insurance, joint stock companies, and other capitalist institutions in the financial centres of Europe and America. This concentration of wealth was based on the appropriation of the technological advances, whatever their origins, in the interests of entrepreneurs who found ways to reap undue profits through activities that relied on theft and slavery.

Selected Readings

- Hamady Bocoum, ed., The Origins of Iron Metallurgy in Africa: New Light on its Antiquity: West and Central Africa (Paris: UNESCO, 2004)
- Judith Ann Carney, Black Rice: the African origins of rice cultivation in the Americas(Cambridge: Harvard University Press, 2001)
- Judith A. Carney, "'With Grains in Her Hair': Rice in Colonial Brazil," Slavery and Abolition, 25, 1 (2004)
- Judith Carney, "Landscapes of Technology Transfer: Rice Cultivation and African Continuities," Technology and Culture, 37, 1 (1996)
- Judith A. Carney and Richard Nicholas Rosomoff, In the Shadow of Slavery: Africa's Botanical Legacy in the Atlantic World (Berkeley and Los Angeles: University of California Press, 2010)
- Crosby, Alfred, The Columbian Exchange: Biological and Cultural Consequences of 1492 (Westport, CN: Greenwood, 1972)
- David Eltis, Philip Morgan and David Richardson, "Agency and Diaspora in Atlantic History: Reassessing the African Contribution to Rice Cultivation in the Americas," American Historical Review, 112,5 (2007), 1329-1358
- Edda L. Fields-Black, Deep Roots: Rice Farmers in West Africa and the African DiasporaBloomington: Indiana University Press, 2008
- Michael A. Gomez, Exchanging Our Country Marks: The Transformation of African Identities in the Colonial and Antebellum South (Chapel Hill: University of North Carolina Press, 1998)
- Gwendolyn Midlo Hall, Slavery and African Ethnicities in the Americas: Restoring the Links (Chapel Hill: University of North Carolina Press, 2005)
- Colleen E. Kriger. Cloth in West Africa History (Lanham, MD: Altamira Press, 2006)
- Colleen E. Kriger. "Guinea Cloth: Production and Consumption of Cotton Textiles in West Africa before and during the

Atlantic Slave Trade," in Giorgio Riello and Prasannah Parthasarathi (eds.), The Spinning World a Global History of Cotton Textiles, 1200-1850(New York: Oxford University Press, 2009)
- Paul E. Lovejoy, "Kola in the History of West Africa," Cahiers d'études africaines, 20, 1/2 (1980), 173-75
- Paul E. Lovejoy, "The 'Coffee' of the Sudan: Consumption of Kola Nuts in the Sokoto Caliphate in the Nineteenth Century," in Jordan Goodman,
- Paul E. Lovejoy, and Andrew Sherratt, eds., Consuming Habits: Drugs in History and Anthropology, London, Routledge, 2nded., 2008), 98-120
- Paul E. Lovejoy, Salt of the Desert Sun. A History of Salt Production and Trade in the Central Sudan (Cambridge University Press, 1986)
- William S. Pollitzer, The Gullah People and Their African Heritage (Athens, GA: University of Georgia Press, 1999)
- Stemler, A.B.L., Harlan, J.R. and Dewet, J.M.J. "Caudatum Sorghums and Speakers of Chari-Nile Languages in Africa," Journal of African History, 16, (1975), 161-83
- John Thornton, "Precolonial Africa Industry and the Atlantic Trade, 1500-1800," African Economic History, 19 (1990), 1-19

****

MONTICELLO
A Project of the Robert H. Smith International Center for Jefferson Studies, Jefferson Library | Monticello

Slave Medicine

West African Medicinal Plants

West African slaves brought not only herbal knowledge with them across the Atlantic; they also imported the actual seeds. Some wore necklaces of wild liquorice seeds as a protective amulet. Captains of slaving vessels used native roots to treat fevers that decimated their human cargo. The ships' hellish holds were lined with straw that held the seeds of African grasses and other plants that took root in New

World soil. Moreover, since the West African climate is similar to the climate in the mid-Atlantic region, slaves may have found counterparts to their own plant species. Certainly the herb lore of captive Africans was expanded by contact with Indian tribes as well as by interaction with Europeans. (Fett, p. 63ff) French missionaries in the 17th Century, for example, noted the use of boneset (*Eupatorium perfoliatum*) by the Iroquois. In the 18th Century, both Europeans and their African slaves used it. By the 19th Century, the medical establishment was regularly prescribing it for fevers.

Although the exchange of plant knowledge was cross-cultural, there were distinct differences in the way this knowledge was preserved and applied. Slave remedies were transmitted orally from generation to generation whereas white domestic healers like Lucy tended to write down cures, along with recipes for preserves and meat pies, in "receipt" books. Their healing recipes were often complex and required considerable amounts of time to prepare. A recipe for dysentery, for example, demanded boiling a "teacup of logwood chips in a pint of sweet milk", then boiling another ten minutes with sugar, and finally burning "four tablespoons of brandy" in a plate, then stirring it into the milk mixture. (Clinton, p. 145) Slave remedies tended to be simpler than white medicines. Teas or poultices were made with one or two plants. Scholar Sarah Mitchell Cotton speculates that slaves had less time to gather ingredients and less time to prepare complex mixtures. Other researchers suggest that simplicity may have been a result of a more sophisticated understanding of the properties of each plant. (Fett, p.74)

Religion and Healing

Unlike the science-based medicine of today, 18th century medicine had a religious component for both black and white healers. As Eileen Malone-Brown observes in her essay, "Healthcare During Lucy's Lifetime," some Europeans saw healing plants as a gift from God and many practitioners resorted to prayer as well as herbs. African healers also felt a sacred connection to plants they found in the woods, and they used elements from African religious rituals when they prepared medicines (Fett, p.76ff). Europeans, however, dismissed African spirituality as "superstition" and an indication of a child-like mentality. Lucy may or may not have exchanged remedies with slave "root doctors" and midwives, but as a devout Christian, she certainly would have disapproved of slave "conjurers" who, in the tradition of their forebears in Africa, cast spells and, along with plants and animal parts, used trickery and intimidation to treat illness of both body and soul. Lucy was a Methodist and the Methodists were known for evangelizing amid slave communities. So it is probable that many of Lucy's slaves, perhaps without altogether abandoning ancestral religious beliefs, adopted their owner's creed. Converted slaves likely would have gained not only her favor but also inspiration from Biblical stories like Exodus that hold the promise of triumph over oppression. (For more on Lucy and slavery, see "Lucy Meriwether Lewis Marks: Her Life and Her World".)

Slave Remedies and Recipes

Slaves used many of the plants used by the community of their white owners: snakeroot, **mayapple**, **red pepper**, boneset, pine needles, comfrey, and red oak bark, to name a few. Slave healers understood the various preparations of **pokeweed** and how to avoid its dangers while taking advantage of its curative properties. **Sassafras** root tea was a popular seasonal blood cleanser believed to "search de blood" for what was wrong and go to work on it. (Fett, p.75). **Jimsonweed** was used for rheumatism, chestnut leaf tea for asthma, mint and cow manure tea "fur consumption". (Genovese, p.169) Slave midwives would have known and used herbs for "female complaints" and to ease childbirth. Slaves preferred their own doctors to white doctors and their "heroic" purging and bloodletting. For an enlightening look at what it meant to be sick when you didn't own your own body, see historian Todd L. Savitt's on-line essay "Slave Health and Medicine: If You Got Sick And Were Black".

Slave Health and Plantation Productivity

The health of a planter's work force was critical to economic success. All slave illnesses had to be reported to a farm's overseer or owner, under pain of punishment. The responsibility for the health of slaves often fell to the mistress of the plantation. In 1781, a year after Lucy married her second husband John Marks, a French nobleman visited William Byrd's large plantation on the James River and remarked that Mrs. Byrd took "great care of her negros, makes them happy as their situation will permit, and serves them herself in times of sickness." He goes on to say:

She has even made some interesting discoveries on the disorders to them, and discovered a very salutary method of treating a sort of putrid fever which carries them off commonly in a few days, and against which the physicians of the country have exerted themselves without avail. (Blanton, p. 169)

Like Mrs. Byrd, Lucy would have spent considerable time attending to her slaves' health, especially since after the death of her second husband, she was not only mistress of her plantation, she was master as well. Presumeably, like her slave-owning neighbor Thomas Jefferson, she would have followed common health practices like providing warm clothing in cold weather and restricting heavy labor for pregnant women at the end of their terms.

Still, despite her investment in the health of her slaves, they would have suffered from complaints ranging from worms and skin ulcers to respiratory diseases and influenzas. (Savitt, p.49ff) Moreover, any complaint could become a point of conflict. It was not simply a question of slaves preferring their own doctors and often being reluctant to report illness to an overseer or owner. There was the more complicated question of malingering. Slaves often complained of sickness to avoid work – a strategy that encouraged a perception of slaves as "lazy" and "deceitful." However, in West African mythology, the Trickster Eshu was a hero. On American shores, he wore the guise of Br'er Rabbit, the admirable scoundrel of the Uncle Remus stories.

Like her fellow slave-owning planters, Lucy would have had to determine whether a slave was actually ill or was faking illness. Given that she was a disciplinarian with her own children, we can suppose she was strict with her "servants" – as she and her peers euphemistically referred to their slaves. Nonetheless, a slave's claim of sickness would have put her on the horns of a dilemma: to ignore a real illness might result in a prolonged, more serious illness or even death. On the other hand, to allow a slave to take to bed reduced her labor force. In either case, farm profits would suffer. (Fett, p.177ff)

Some plantations like George Washington's Mount Vernon which had over 300 slaves had separate quarters for sick slaves. We do not know if Lucy, who owned 47 slaves at the time of her death, had a "sickhouse" at Locust Hill . No doubt she was assisted in her doctoring by slave nurses who, under her supervision, would have administered medicines, prepared gruels. and cleaned patients and their bedclothes of blood, vomit and excrement .(Fett. p.118) We do know that Lucy paid a neighbor, a Mrs. Via to "wait on Frankie", a pregnant female slave. Lucy may also have owned a slave "doctoress" or "granny" who would have combined her herbal knowledge with midwifery skills. (Blanton, 174) Slave midwives were more valuable and earned their owners (and perhaps themselves) money by safely delivering slave babies, and in some cases, white babies as well.

Legal Restrictions on Slave Doctors

If healing skills were an advantage to both slave and owner, they also posed a threat. Slaves who knew their plants had easy access to poison. They knew which herbs could abort pregnancies, which could sicken, which could cause sudden death. In Virginia, the decades between 1791 and 1809 were marked by a series of slave uprisings; the most notable of these was the Gabriel Plot in which over a thousand slaves attacked the city of Richmond. In 1831, Nat Turner led his famous slave rebellion in Virginia's Tidewater. Virginia state records show that between 1780 and 1864, 58 slaves were convicted of poisoning or attempted poisoning. (Blanton, p. 174) No matter how much white slave owners relied on and trusted slave healers, they also feared them.

This fear prompted the enactment of laws. As early as 1748, the colony of Virginia forbade "any negroe, or other slave" to administer "any medicine whatsoever" under pain of death "without benefit of clergy". An exception was made for slaves treating other slaves or her owner's family, providing the owner gave permission. In 1792, the law was softened to allow acquittal if, at the slave practitioner's trial, it was shown that there had been "no ill intent and no bad consequence." Thus, if a "doctoress" administered a preparation of herbs and the patient happened to die, an angry owner could have the slave hanged. As an earlier historian observed: "With stringent laws and with fear of poisoning constantly in the public mind, it is surprising that any of the negroes should have attempted the risky business of prescribing for the sick." (Blanton, p. 174)

In fact, the laws were no deterrent. Slave medicine flourished on plantations. While collecting wild herbs and roots, slave doctors, male and female, escaped the boundaries of their working life and perhaps experienced a fleeting taste of physical freedom. Certainly, a belief in the sacredness of healing plants allowed them to connect with an authority higher than their owners – be it animistic African dieties or a single Christian god. In treating fellow slaves, they became an instrument of divine power. They, not their owner, controlled a patient's body. At its core, slave healing was an empowerment for both healer and patient.

Christine Andreae

Sources:

Aptheker, Herbert, *American Negro Slave Revolts*. New York: International Publishers. 1970

Blanton, *Medicine in Virginia in the 18th Century*

Clinton, *The Plantation Mistress*

Fett, Sharla M. *Working Cures: Healing, Health, and Power on Southern Plantations*. Chapel Hill and London: University of North Carolina Press, 2002

Genovese, Elizabeth Fox, *Within the Plantation Household*

Savitt, Todd L., *Medicine and Slavery*

Stanton, Lucia. *Slavery at Monticello*. Thomas Jefferson Foundation, Monticello Monograph Series, 1996

Web source:

Cotton, Sarah Mitchell: *Bodies of Knowledge: The Influence of Slaves on the Antebellum Medical Community* at

http://scholar.lib.vt.edu/theses/available/etd-65172149731401/unrestricted/CH1.PDF

****

Historians expose early scientists' debt to the slave trade

By Sam Kean Apr. 4, 2019

At the dawn of the 1700s, European science seemed poised to conquer all of nature. Isaac Newton had recently published his monumental theory of gravity. Telescopes were opening up the heavens to study, and Robert Hooke and Antonie van

Leeuwenhoek's microscopes were doing the same for the miniature world. Fantastic new plants and animals were pouring in from Asia and the Americas. But one of the most important scientists alive then was someone few people have ever heard of, an apothecary and naturalist named James Petiver. And he was important for a startling reason: He had good connections within the slave trade.

Although he rarely left London, Petiver ran a global network of dozens of ship surgeons and captains who collected animal and plant specimens for him in far-flung colonies. Petiver set up a museum and research center with those specimens, and he and visiting scientists wrote papers that other naturalists (including Carl Linnaeus, the father of taxonomy) drew on. Between one-quarter and one-third of Petiver's collectors worked in the slave trade, largely because he had no other options: Few ships outside the slave trade traveled to key points in Africa and Latin America. Petiver eventually amassed the largest natural history collection in the world, and it never would have happened without slavery.

Strikingly, some naturalists also instructed their contacts abroad to train slaves as collectors. Slaves often knew about specimens that Europeans didn't and visited areas that Europeans wouldn't. Those slaves virtually never got credit for their work, though Petiver did offer to pay them a half-crown ($18 today) for every dozen insects or 12 pence ($7) for every dozen plants.

James Delbourgo, a historian at Rutgers University in New Brunswick, New Jersey, who has written extensively about slavery and science, agrees. He argues that the belief in the progressive nature of science has made historians reluctant to take a critical look at its past. "This is a hard story for us to deal with," he says. He adds that academic specialization also prevented many people from seeing what, in retrospect, seem like obvious connections: "Slave trade historians don't know about science, and vice versa."

Kathleen Murphy, California Polytechnic State University

Those connections aren't just ancient history. Thousands of specimens collected through the slave trade still reside in places such as the Natural History Museum in London, and they're still used in genetic and taxonomic research. Yet few people using the collections know of their origins.

All of which casts an uncomfortable shadow on what's often viewed as a heroic era in science. "We do not often think of the wretched, miserable, and inhuman spaces of slave ships as simultaneously being spaces of natural history," Murphy writes in *The William and Mary Quarterly*. "Yet Petiver's museum suggests that this is exactly what they were."

Compromises to gain access to distant lands

Why did scientists align themselves with that horror? Access. European governments did sometimes sponsor scientific expeditions, but most ships visiting Africa and the Americas were private vessels engaged in the "triangular trade." That three-way exchange sent guns and manufactured goods to Africa; slaves to the Americas; and dyes, drugs, and sugar back to Europe. To gain access to Africa and the Americas, scientists had to hitch rides on slave ships. Upon arrival, the naturalists also relied on slavers for food, shelter, mail, equipment, and local transport.

****

Seeds of Memory: Botanical Legacies of the African Diaspora

J. Carney, Ph.D.
Department of Geography, University of California Los Angeles

The decades following 1492 launched an era of European overseas expansion, which led to an unprecedented intercontinental exchange of plant and animal species. Historian Alfred W. Crosby famously called this process the *Columbian Exchange*.

R. Voeks and J. Rashford (eds.), *African Ethnobotany in the Americas*, 13
DOI 10.1007/978-1-4614-0836-9_2, © Springer Science+Business Media New York 2013

In his second book, ***Ecological Imperialism***, Crosby drew attention to the significance of ordinary people for the transfer of species across the globe. These were European emigrants to new lands—people who were not operating as administrators, scientists, or representatives of colonial institutions such as botanical gardens, scientific societies, and museums. With the plant and animal species that accompanied them, European settlers transformed the environments of Australia, New Zealand, and South Africa into, in Crosby's words, *Neo-Europes* (Crosby 1972, 1986).

However, the Columbian Exchange literature ignores another important intercontinental species transfer that over the same period of time also occurred as a consequence of immigration. In this instance, the migration was not voluntary but forced, and the agents of dissemination involved not just Europeans but also enslaved Africans. Reference is to the African plants and food animals that proved instrumental for European colonization of the New World tropics. The movement of these biota across the Atlantic Ocean to tropical America in the first century of plantation development depended significantly on the transatlantic slave trade for their dispersal. Plants and animals arrived on slave ships together with African captives for whom they were traditional dietary staples, medicinals, and food animals. Although the Columbian Exchange celebrates the role of New World and Asian crops in revolutionizing food systems of Africa, there is little attention given to the impact of African species in lowland tropical America. Of interest is the contrasting significance of these African species for slaveholders and the enslaved.

Canada

From Library and Archives Canada (https://www.bac-lac.gc.ca/eng/discover/immigration/history-ethnic-cultural/Pages/blacks.aspx):

There has been a steady stream of migration of Black people into Canada via Africa, Europe, the Caribbean, and the United States since the 17th century. The first recorded Black person to arrive in Canada was an African named Mathieu de Coste who arrived in 1608 to serve as interpreter of the Mi'kmaq language to the governor of Acadia. A few thousand Africans arrived in Canada in the 17th and 18th centuries as slaves. After the American Revolution, the British gave passage to over 3000 slaves and free Blacks who had remained loyal to the Crown. These Black Loyalists joined the many other United Empire Loyalists in settlements across the Maritime Provinces of Nova Scotia, New Brunswick, and Prince Edward Island. Other Black slaves joined their Loyalist slave owners when they migrated to Canada.

In 1793, the Upper Canada legislature passed an act that granted gradual abolition and any slave arriving in the province was automatically declared free. Fearing for their safety in the United States after the passage of the first *Fugitive Slave Law* in 1793, over 30,000 slaves came to Canada via the Underground Railroad until the end of the American Civil War in 1865. They settled mostly in southern Ontario, but some also settled in Quebec and Nova Scotia. Many returned to the United States to fight in the Civil War and rejoin their families after its end.

Other migrations of Black people from the United States occurred during the War of 1812, when over 2000 refugees came to Nova Scotia and New Brunswick. Another group of over 800 free Blacks from California migrated to Vancouver Island between 1858 and 1860. Many Black people migrated to Canada in search of work and became porters with the railroad companies in Ontario, Quebec, and the Western provinces or worked in mines in the Maritimes. Between 1909 and 1911 over 1500 migrated from Oklahoma as farmers and moved to Manitoba, Saskatchewan, and Alberta.

****

Throughout the 1800's, a number of historic Black communities were established across Canada. Some of these communities came as a result of war. Also, between 1800 and 1865, approximately 30,000 Black people came to Canada via the Underground Railway – the network of secret routes and safe houses used by enslaved Africans to escape into free American states and Canada with the support of abolitionists and their allies. (Historic Black Canadian communities - Canada.ca)

****

Anderson Ruffin Abbott: First Canadian-born Black doctor in Canada (1837-1913)

In 1861, Abbott became the first Canadian-born man of Black heritage to become a licensed physician. He worked as one of eight black surgeons during the American Civil War. Abbott returned to Canada in 1866, where he established a medical practice. Admitted to the College of Physicians and Surgeons of Ontario in 1871

William Peyton Hubbard: Canadian inventor and politician 1842-1935

Born in Toronto, William Peyton Hubbard was the son of American slaves who escaped their plantation in Virginia and came to Canada via the underground railroad. Hubbard was trained as a baker at the Toronto Normal School; the façade of the building is now located on the Ryerson University campus.

Building on his training, Hubbard invented and patented a commercial baker's oven – the Hubbard Portable – which he sold through his company, Hubbard Ovens.

Elijah McCoy (1843 – 1929): Canadian-American inventor and engineer

Born in Colchester, Ontario, Elijah McCoy's parents were former slaves who had escaped to Canada using the underground railroad. He was educated in black schools in the Colchester Township, due to the 1850 Common Schools Act which segregated the Upper Canada schools in 1850. At age 15, Elijah McCoy was sent to Scotland for an apprenticeship and study, and was certified as a mechanical engineer.

In 1872, he invented an automatic lubricator for locomotive steam engines and obtained his first patent.

United States

Extraordinary powers of calculation

Thomas Fuller, known as the Virginia Calculator, was kidnapped from his native Africa at the age of fourteen and sold to a planter.

Challenge
He was regularly tested. On being asked, " How many seconds a man has lived who is 70 years, 17 days and 12 hours old?" He answered in a minute and a half 2 210 500 800. One of his examiners who was using pen and paper, advise him his answer was incorrect! The sum was not so great as he had suggested. Upon which Thomas hastily replied: "Stop, you forget all the leap years!" On adding the amount of the seconds for the leap years, the total in both their sums agreed exactly.

Achievements
Later in his life was he was discovered by antislavery campaigners, who used him as a role model to demonstrate that blacks were not mentally inferior to whites.

Benjamin Banneker

Benjamin Banneker was a largely self-educated mathematician, astronomer and first civil rights leader.

Challenge

He would exchanged letters with the soon to be Third U.S. President Thomas Jefferson, whom had stated that Negroes are inferior to whites when it comes to high level of mathematical thinking.

Achievements
Banneker was the first scientist to study the relativity of time and space, and his revelations on the topic preceded Einstein's Theory of Relativity by two centuries. He successfully predicted the solar eclipse that occurred on April 14, 1789, contradicting the forecasts of prominent mathematicians and astronomers of the day. Banneker was the first to disclose in his writings that the Star of Sirius is two stars rather than one. His hypothesis was not confirmed until the event of the Hubble Telescope two and a half centuries later at NASA.

Thomas Jennings

He was the first African American to receive a patent. He earned his patent in 1821, at age 30, for a process that is the forerunner of modern dry-cleaning. Jennings was a freeman who was born in the United States. He was an accomplished tailor whose reputation enabled him to open up his own clothing store in New York. Along the way, he devised his patented dry-scouring process. A successful businessman, Jennings was a financial supporter of the abolitionist cause.

Henry Blair

Presumably a freeman, is the second African American to receive a patent. His two harvester patents—granted in 1834 and 1836—are the only ones ever to indicate that they were granted to a "colored man."

Latin America

The idea of Afro-Latin America was introduced to the United States by political scientists Anani Dzidzienyo and Pierre-Michel Fontaine, both of whom were doing research in Brazil on black social and political movements. Dzidzienyo published his findings in a 1978 article on "Activity and Inactivity in the Politics of Afro-Latin America"; two years later Fontaine reported on "The Political Economy of Afro-Latin America." Fontaine defined the term to include "all regions of Latin America where significant groups of people of known African ancestry are found." But that definition left at least two questions unanswered. First, how do we "know" when people are of African ancestry? And second, how large do groups of those people need to be before we consider them "significant"? (from Revista, Harvard Review of Latin America, https://revista.drclas.harvard.edu/book/afro-latin-america-numbers-politics-census).

****

Between the 1490s and the 1850s, Latin America, including the Spanish-speaking Caribbean and Brazil, imported the largest number of African slaves to the New World, generating the single-greatest concentration of black populations outside of the African continent. This pivotal moment in the transfer of African peoples was also a transformational time during which the interrelationships among blacks, Native Americans, and whites produced the essential cultural and demographic framework that would define the region for centuries. (**The Black Experience in Colonial Latin America**, by Ben Vinson, Greg Graves, III)

****

From Afro-Latin America, 1899 – 2000, George Reid Andrews, Research Professor of History

Elided in the USA census is that "blacks" and "Hispanics" are not neceassarily separate groups. In the nations of latin America, people of African ancestry are an estimated one-quarter of the total Black population. Indeed, the heart of the New World African Diaspora lies not north of the border, in the United States, but south. During the period of slavery, ten times as many Africans came to Spanish and Portuguese America as to the United States. By the end of the 1900s, Afro-Latin Americans outnumbered Afro-North Americans by three to one, and formed, on average, almost twice as large a proportion of their respective populations. (pg 3)

****

Many voluntary Black immigrants to the United States do not refer to themselves as Africa-Americans; but accept being called Black Americans. So the census in the United States should use the term "Black Americans when referring to Black people as a whole.

****

Afro-Latin America, 1800. Numbers under country names indicate the size of the black and brown (pardo) population, in 000s. Credit: William Nelson

Afro-Latin America, 1900. Numbers under country names indicate the size of the black and brown (pardo) population, in 000s. Credit: William Nelson

Afro-Latin America, 2000. Numbers under country names indicate the size of the black and brown (pardo) popu-lation, in 000s. . Credit: William Nelson

Afro-Latin America, 2010. Numbers under country names indicate the size of the black and brown (pardo) popu-lation, in 000s. (Map by Lena Andrews.)

Afro-descendants in Latin America
In thousands

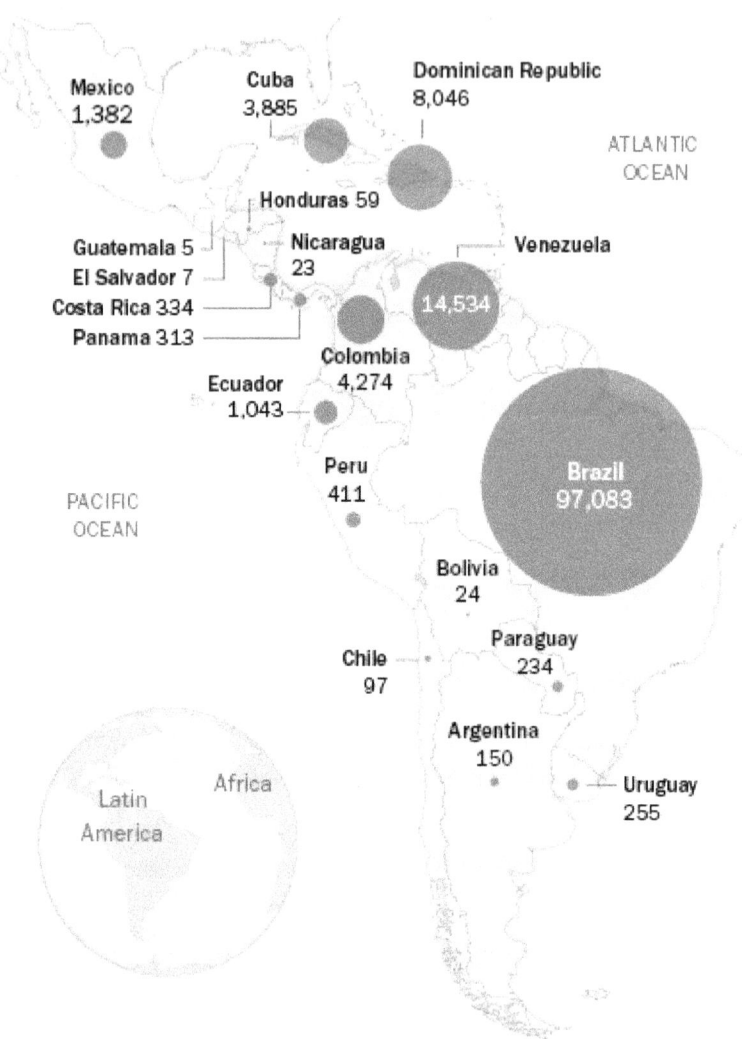

Note: Estimates for the Dominican Republic include the term "indio" and estimates for Venezuela include the term "moreno" as afro-descendant.
Source: For Mexico: 2015 data from Mexico's Instituto Nacional de Estadística y Geografía, INEGI. For all other countries: Edward Telles and the Project on Ethnicity and Race in Latin America (PERLA), "Pigmentocracies: Ethnicity, Race, and Color in Latin America" (2014).

PEW RESEARCH CENTER

….by 1600, the coastal regions of Bahia and Pernambuco accounted for over one-half of the world's sugar production. (pg 13)

Slaves also labored in a variety of urban occupations in Brazil, Argentina and Peru, ranging from the most unskilled and degraded to the most highly skilled. They were prominent in any enterprise requiring large groups of laborers gathered in one place, suchas construction and manufacturing, food processing establishments, such as bakeries, or the meat salting and drying factories………..(pg 15).

The intensified imports of Africans occurred due to economic expansion in Europe.

In 1789, Spain removed all restrictions on the slave trade to its colonies. Ships flying the flag of any nation were now permitted to bring slaves into Spanish ports. This policy led to ………..the Caribbean became the new center of sugar production. (pgs 18 -19).

****

Caribbean

From the *British library Collection:*

The Europeans came to the Caribbean in search of wealth. The Spanish had originally looked for gold and silver, but there was little to be found. Instead, the Europeans tried growing different crops to be sold back home. After unsuccessful experiments with growing tobacco, the English colonists tried growing sugarcane in the Caribbean. This was not a local plant, but it grew well after its introduction. Sugarcane could be used to make various products. There was sugar, of course, which went well with tea, coffee and chocolate. It could also be used to make rum, a strong alcoholic drink. A lot of people in Europe wanted such products, and, as a result, those who grew it – known as 'planters' – became very wealthy. This also made the Caribbean colonies valuable – and tempting targets for rival empires. Britain and France were constantly at war in the 18th century and early 19th century, with places such as Martinique changing hands many times.

****

From ***The National Archives***, We are a non-ministerial department, and the official archive and publisher for the UK Government, and for England and Wales. We are the guardians of over 1,000 years of iconic national documents:

Between 1662 and 1807 Britain shipped 3.1 million Africans across the Atlantic Ocean in the Transatlantic Slave Trade. Africans were forcibly brought to British owned colonies in the Caribbean and sold as slaves to work on plantations. Those engaged in the trade were driven by the huge financial gain to be made, both in the Caribbean and at home in Britain.

Enslaved people constantly rebelled against slavery right up until emancipation in 1834. Most spectacular were the slave revolts during the 18th and 19th centuries, including: Tacky's rebellion in 1760s Jamaica, the Haitian Revolution (1789), Fedon's 1790s revolution in Grenada, the 1816 Barbados slave revolt led by Bussa, and the major 1831 slave revolt in Jamaica led by Sam Sharpe. Also voices of dissent began emerging in Britain, highlighting the poor conditions of enslaved people. Whilst the Abolition movement was growing, so was the opposition by those with financial interests in the Caribbean.

The British slave trade officially ended in 1807, making the buying and selling of slaves from Africa illegal; however, slavery itself had not ended. It was not until 1 August 1834 that slavery ended in the British Caribbean following legislation passed the previous year. This was followed by a period of apprenticeship with freedom coming in 1838.

Even after the end of slavery and apprenticeship the Caribbean was not totally free. Former enslaved people received no compensation and had limited representation in the legislatures. Indentured labour from India and China was introduced after slavery. This system resulted in much abuse and was not abolished until the early part of the 20th century. After indenture, Indians and Africans struggled to own land and create their own communities.

****

The Franch lost to Black Haitians.

Louisiana Purchase, the bridge for USA between East ansd West.

In 1804, after 13 years of war, Haiti, with its overwhelmingly Black population and leadership, became the second nation in the Americas to win independence from a European state when the army of former slaves defeated Napoleon's invasion force. During the 19th Century, further waves of rebellion, such as the Baptist War, led by Sam Sharpe in Jamaica created the conditions for the incremental abolition of slavery in the region, with Cuba the last island to achieve emancipation in 1886.

****

Caribbean Science Foundation *(CSF)*

Europe

From *Black Africans in Renaissance Europe*, edited by T.F. Earle and K.J.P. Lowe

No such place as generic "black Africa" existed or exists; Africa was/is a vast continent, full of cultural, social, religious, linguistic and ethnic diversity, and of regional difference. But the process of removing Africans to Europe in the Renaissance period served to rob them of these distinguishing features, taking away their old, nuance identities and providing them instead with new, one dimensional European ones labeling them all as "black Africans". Arricval in Europe as slaves meant the systematic erasure of all the more significant aspects of their past, starting with their names, their languages, their religions, their families and communities, and their culturalpractices, but it did not erase their appearance. Hence the use of the tern "black Africans' and, in order to maintain parity of terminology, the use of the similarly non-existent construct 'Renaissance Europeans'. (pg 2)

****

Asia

Slaves were a heterogeneous group but this is not surprising because there were several waves of migration at different time periods involving various slavers. Many slaves had originated in Mozambique and Madagascar. But there were also a few Indonesian slaves from the islands of Nias and Batta, off the west coast of Sumatra, illustrating that there were both Asian and African slaves working for the Company.

Africans were moved to Macau (China) and Nagasaki (Japan) and even Timor. They were also taken to Peru, Mexico and Colombia. The global movement of Africans should be recognised. Public understanding (History and Policy, Cambridge, UK)of the African diaspora is inevitably influenced by the transatlantic slave trade

****

From Slavery & Abolition, A Journal of Slave and Post-Slave Studies, Vol. 38, 2017 – Issue 1

Throughout the seventeenth and eighteenth centuries, the slave trade in Asia was conducted by both European and Asian traders. Gujarati merchants transported enslaved Mozambicans to Daman and Diu.[24] The two major players in Maluku trade, including slavery, were the Chinese and the Bugis.[25] As in Maluku, Chinese slave traders were the major players in the export of Balinese slaves, mostly to Batavia, in the seventeenth and eighteenth centuries.[26] Indian slaves were exported by European powers to their colonies in the eastern and western Indian Ocean colonies.[27] High-ranking VOC (The Dutch East India Company, officially the United East India Company) officials – merchants, administrators and governors,– were sometimes able to fit out their own private ships to engage in the slave trade. Lower- and middle-ranked personnel employed in intra-Asiatic shipping, used the opportunity of their voyages in Asia on board Company ships to transport slaves.

****

The African Diaspora in the Indian Ocean World
by: Omar H. Ali

Abstract

Over the course of nearly 20 centuries, millions of East Africans crossed the Indian Ocean and its several seas and adjoining bodies of water in their journey to distant lands, from Arabia and Iraq to India and Sri Lanka.

Called Kaffir, Siddi, Habshi, or Zanji, these men, women and children from Sudan in the north to Mozambique in the south Africanized the Indian Ocean world and helped shape the societies they entered and made their own.

Free or enslaved, soldiers, servants, sailors, merchants, mystics, musicians, commanders, nurses, or founders of dynasties, they contributed their cultures, talents, skills and labor to their new world, as millions of their descendants continue to do. Yet, their heroic odyssey remains little known.

The African Diaspora in the Indian Ocean World traces a truly unique and fascinating story of struggles and achievements across a variety of societies, cultures, religions, languages and times.

****

African Migrants as cultural brokers in South Asia Shihan de Silva Jayasuriya, FRAS Institute of Commonwealth Studies, University of London

The bicentennial of the abolition of the Atlantic slave trade has obscured other slave routes. African slaves reached not only the Caribbean but also China and Japan, spilling over into the Pacific Ocean. The process of migration eastwards began long before the Atlantic slave trade. While the latter lasted mainly from about 1440 to 1870, the movement across the Red Sea and the Indian Ocean spanned over several millennia (de Silva Jayasuriya 2006a). This general ignorance regarding the more ancient migration is partly explained by difficulties in identifying Africans in Asia (de Silva Jayasuriya 2006b).

Generally Africans were marginalized. They have therefore lacked a voice but made their presence felt through a series of distinct cultural expressions. Africans migrated into circumstances and societies that varied. This paper cannot explore all the African spaces in South Asia but a few examples will demonstrate the role of migrants as cultural brokers between the two continents of Africa and Asia.

The term Diaspora is contested and open to academic debate. The reality is that descendants of Africans are born and live in Asian countries which have become their homes today. In most cases they do not have another home. Their physiognomic features are African. Chanting, controlled breathing and rhythmic bodily movements leading to a trance and spirit possessions were common practice in East African coastal towns and even on the dhows on which the slaves were taken to be sold. Musical traditions are one indicator of the African presence in Asia. A study of these traditions and the lyrics can forge links between Africa and Asia.

Goma and Laywa: India and Pakistan

Africans have been called by various names in South Asia. As their migration to India is an old phenomenon, the terms describing them have also changed over time and space. The word Africa is a 20th century term. Previously the blacks were perceived to have come from Sudan, Habasha, Zandj or Nuba. People of African descent in South Asia have mostly been referred to as Habshi, Kaffir and Sidi. All these words have Arabic etyma. The best known were the Habasha (Ethiopians), since they were geographically closer to Arabia and associated with Prophet Muhammad. Bilal, the first Muezzin was the son of an Ethiopian slave. Kaffir is from the Arabic word qafr meaning 'non-believer' and was originally used by Muslims to refer to the 'non-muslims'. The etymon of the word Sidi lies in Arabic Seyidi/Sayeedi/Sayedi meaning 'lord or master'.

Today the largest Afro-Indian communities are spread over several States of India but mainly in Gujarat, Karnataka and Andhra Pradesh. Smaller communities are found in Maharashtra, Madhya Pradesh, Tamil Nadu and Uttar Pradesh (de Silva Jayasuriya 2007a, 2008b). Afro-Gujaratis have been performing in Europe, America and Africa since October 2002.They play sacred music and dance, singing to their ancestral Saint, Bava Gor, who is believed to have been an Abyssinian. They perform dhamal which they call goma, a word whose etymon is found in a Swahili word ngoma meaning 'drum' and also 'dance'.

Bibliography

Alpers, E. "The African Diaspora in the Indian Ocean: A Comparative Perspective", in « The African Diaspora in the Indian

Ocean, eds: Shihan de S Jayasuriya & Richard Pankhurst, New Jersey: Africa World Press, Pp.19-52, 2003.

Badalkhan, S. "On the Presence of African Musical Culture in Coastal Balochistan", in « Journeys and Dwellings: Indian Ocean Themes in South Asia, ed. H. Basu. New Delhi: Orient Longman, 2006.

Basu, H. "The Sidi and the Cult of Bava Gor in Gujarat", « Journal of Indian Anthropological Society 28 (1993), pp 289-300.

Christensen, D. "Musical Life in Sohar, Oman" in « The Garland Encyclopedia of World Music: The Middle East, eds. V Danielson, S Marcus and D Reynolds, New York: Routeledge, 2002, pp. 671- 683.

De Silva Jayasuriya, S. The African Presence in Sri Lanka, in « The African Diaspora in the Indian Ocean. eds. S de Silva Jayasuriya & R Pankhurst. New Jersey: Africa World Press; "Trading on a Thalassic Network: African Migrations Across the Indian Ocean", in « International Social Sciences Journal, pp. 215-225, UNESCO, Paris (2006a); "Identifying Africans in Asia: What's in a Name?" in « African and Asian Studies 5, nos. 3-4, 275-303 (2006b); "Music and Memories: Oral Traditions from an Indian Ocean Island", ed. S de Silva Jayasuriya, MUSIKE, 3 (2006c); "The African Diaspora in India", in « An Encyclopedia of the African Diaspora. Ed: C Boyce-Davies. California: ABC CLIO. (2007a); "The Indian Oceanic African Diaspora" in « An Encyclopedia of the African Diaspora. Ed: C Boyce-Davies. California: ABC CLIO. (2007b); "Migrants and the Maldives: African Connections" in « Uncovering the History of Africans in Asia. Eds. S de Silva Jayasuriya & J-P Angenot (2008a); "African Identity in Asia". Princeton, New Jersey: Markus Wiener Publishers (2008b); "The Portuguese in the East: A Cultural History of a Maritime Trading Empire", London: I.B. Tauris.

Forbes, A. & Ali, F, "The Maldive Islands and their Historical Links with the Coast of Eastern Africa" in "Kenya Past and Present 2", 1980.

Kassebaum, G. and Clause, P. 2000. "Karnataka" in « The Garland Encyclopedia of World Music, ed. Alison Arnold. New York: Routeledge. pp. 866-888.

Minda, A. Personal communication, 2005.

Mohamed, N. "Essays on Early Maldives", National Centre for Linguistic and Historical Research, Male, 2006.

Robbins, K. & McLeod, J.: "African Elites in India", Mapin Publishers, Hyderabad, 2006.

****

China

From BlackPast:

Beginning with the Tang dynasty (618 A.D. to 907 A.D.) documented evidence of contact and trade exists showing a relationship between China and the city-states of east Africa. This relationship has evolved over the centuries and led to a migration of Africans to China to study, trade, and act as diplomats. At least one account indicates that Du Huan was the first Chinese to visit Africa, probably in Nubia, during the 8th century A.D.

By the time of the Ming Dynasty (1368 A.D. to 1644 A.D.) there was extensive trade between the Chinese and the east African city-states of Mogadishu, Malindi, and Kilwa in the modern nations of Somalia, Kenya, and Tanzania respectively. The Chinese imported ivory, rhinoceros horn, amber, and exotic animals such as zebras, ostriches, and giraffes from east Africa. In turn, the city-states received silk, porcelain, and lacquer.

Over the next four centuries the rise of Europe and in particular European trade and colonial expansion marginalized Chinese-East African contact. Both the Chinese and the Africans now looked to Europe and the West rather than each other as trading partners. Trading connections between China and East Africa were not lost but neither were they considered particularly important in this new era of global commerce.

The first significant African American contact with "modern" China came during the Boxer Rebellion. Troops from the 10th Cavalry, one of the four famed *Buffalo Soldier* units, were part of the international military force of 20,000 soldiers sent to suppress the uprising led by the Society of the Righteous and Harmonious Fists (Boxers) and to free foreign hostages and Chinese Christians held by them.

Despite the tensions among university students, Africans and increasingly African Americans have been a growing presence in the country since 1990. By 2014 an estimated 500,000 Africans, Afro-

Caribbeans, and African-Americans were present in China. In comparison there are about one million Chinese living in Africa.

From The British Museum blog:

Chinese exports to Africa

Glazed pottery such as this distinctive jade-green coloured bowl was popular in parts of Africa and notably used as table ware. Made in China during the Ming dynasty (1368–1644) this bowl was found in Malindi, on the coast of Kenya.

African and Southeast Asian textile traditions

The tradition of making batik is well-known in various Southeast Asian countries. This art form of applying a wax resist prior to dyeing fabric is also very popular in West Africa, where it has influenced a variety of cultures and inspired new motifs. There is a similar cloth in the Museum's collection of African textiles, which has design motifs inspired by this Javanese batik for African markets. If you compare the two cloths, you can see many similarities in design.

****

Japan

In the mid-16th century, black Africans arrived in Japan alongside white Europeans, as crew members and slaves.

African samurai: The enduring legacy of a black warrior in feudal Japan

Yasuke was a slave turned samurai from Africa who lived in Japan in the sixteenth century. Yasuke, possibly from Mozambique, arrived in Japan in the late-16th century alongside Jesuit missionary Alessandro Valignano. He found favor with Oda Nobunaga, the Daimyō and warlord, and ultimately became a samurai.

Today, Yasuke's legacy as the world's first African samurai is well known in Japan, spawning everything from prize-winning children's books to a manga series titled "Afro Samurai."
Yasuke's origins remain a mystery as historical sources are scant. While some researchers believe he was from Mozambique, others suggest Sudan.

According to Gary Leupp, a professor of history at Tufts University, Yasuke was taken prisoner by Oda's enemies but later released because he was not Japanese. Yasuke had become a "ronin" -- a samurai without a master.

****

The African Diaspora in Asian Trade Routes and Cultural Memories
Shihan de Silva Jayasuriya | May 2011

Africans were moved to Macau (China) and Nagasaki (Japan) and even Timor.

****

From The Schomburg Center for Research in Black Culture:

Pakistan

Many of the Africans brought into the Indian subcontinent entered through the ports of Baluchistan and Sindh, where they worked as dockworkers, horse-keepers, domestic servants, agricultural workers, nurses, palanquin carriers and apprentices to blacksmiths and carpenters.

Pakistan has the most people of African descent in South Asia. It has been estimated that at least a quarter of the total population of the Makran coast is of African ancestry—that is, at least 250,000 people living on the southern coast of Pakistan, which overlaps with southeastern Iran, can claim East African descent. Beginning in 1650 Oman traded more heavily with the Lamu archipelago on the Swahili coast and transported Africans to the Makran coast. As a result, today many Pakistani of African descent are referred to as Makrani, whether or not they live there.

India

The history of India's Africans, called Siddis, is the best known in the region—largely because of the documentation on those who rose to high positions as military commanders. *Siddis Today*

A number of Siddis converted to Christianity in the 20th century and were sent to Mauritius, the Seychelles and Kenya with support from Christian missionaries. Those who went to Kenya settled in Freretown, near Mombasa. However, they remained relatively isolated, given that the majority of people around them were Muslim.

Today, the number of Siddis in India, who include Muslims, Hindus and Christians, is estimated to be over 50,000.

Sri Lanka and the Maldives

As early as the fifth century, Abyssinians traveled to Sri Lanka (Ceylon) and traded in Matota in the northwest. Centuries later, the Portuguese were the first Europeans to bring Africans to Sri Lanka as slaves and mercenary soldiers. The Portuguese had preceded the Dutch, French and British into the long-existing Indian Ocean trade networks, driving the largely forced migration of Africans into various parts of this world.

Although there is historical evidence that Ethiopians were trading with Sri Lankans in the 5th century, the Afro-Sri Lankans today trace their roots back to the colonial period when the Portuguese, Dutch and British brought African slaves to the Island (de Silva Jayasuriya 2003).

Indoeasia

http://afroeurope.blogspot.com/

Indo-Africans: The forgotten story of the Black Dutchmen

The story of the Black Dutchmen is an almost forgotten part of African, Dutch and Indonesian black History. That's why the Indo-Africans are now exploring their roots.

The Black Dutchmen are the descendants of 3,000 young West African men who were "bought" between 1831 and 1872 by the Dutch colonial army to help crush uprisings in what is now Indonesia.

They were given Dutch nationality, Dutch names, and many of the privileges of the colonial masters.

Many of these men chose to settle in Central Java, and took native wives. Their sons continued to serve in the colonial army until Indonesia's independence. The Belanda Hitam, or Black Dutchmen, as they were called, then sailed off the Netherlands, the homeland they had never seen.

Today every other year or so, the Belanda Hitam, or Black Dutchmen as they are now known in the Netherlands, gather to celebrate their unique ancestry, the RNW reports.

The scene is utterly confusing, even for those most at ease in multicultural settings: tall, black, curly haired older men in African attire; short, fragile, dark skinned, flat nosed women draped in Asian prints; the lighter skinned, more Mediterranean looking youths, and then the lighter haired Caucasians. What they have in common are their Indo-African roots.

With African music, Indonesian food, and an Indo-African fashion show, about 150 of them gathered again recently to share stories about their common ancestors.

"Our story"

The African roots were long kept hidden. Was it shame, or the desire to blend into Dutch society that prevented the Indo-African elders from revealing their secret?

"It's our story", says Joyce Cordus, and it's still largely unknown in the Netherlands", despite several books having been written about the Black Dutchmen. Her father is Daan Cordus, 89, the oldest remaining descendent of the group in the Netherlands who has devoted the past decades to gathering information about their shared background. Today, Joyce is taking over his responsibility as chairman of their association.

Like her, young members of the 5th generation want to understand why they feel different in Dutch society. "They have lots of questions and we have to look for other ways to reach them." Joyce plans to do that by using social media platforms such as Facebook, its Dutch equivalent Hyves, and Twitter.

Interesting to note is that the same indo-black ethnicity is also present in Surinam. A large group of Javanese people from former Dutch colony of Indonesia were brought to Surinam to serve as contract labourers after the abolition of slavery in 1863.

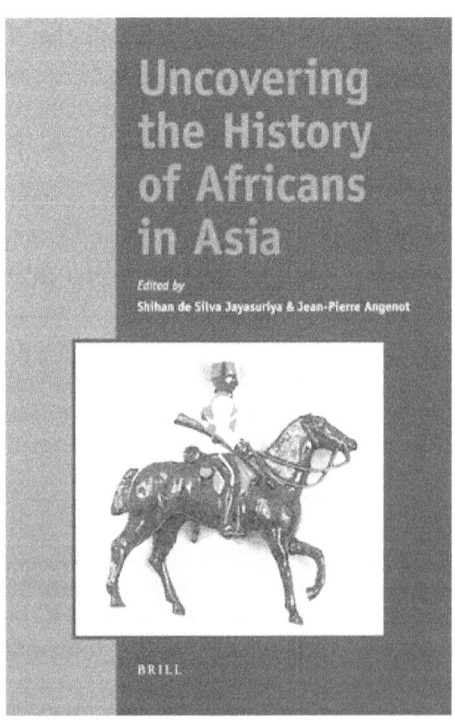

Uncovering the History of Africans in Asia, Editors: Shihan de Silva Jayasuriya and Jean-Pierre Angenot

The BLACKS of PREMODERN CHINA, Don J. Wyatt

Contemporary

(1900 AD – Present)

AFRICA

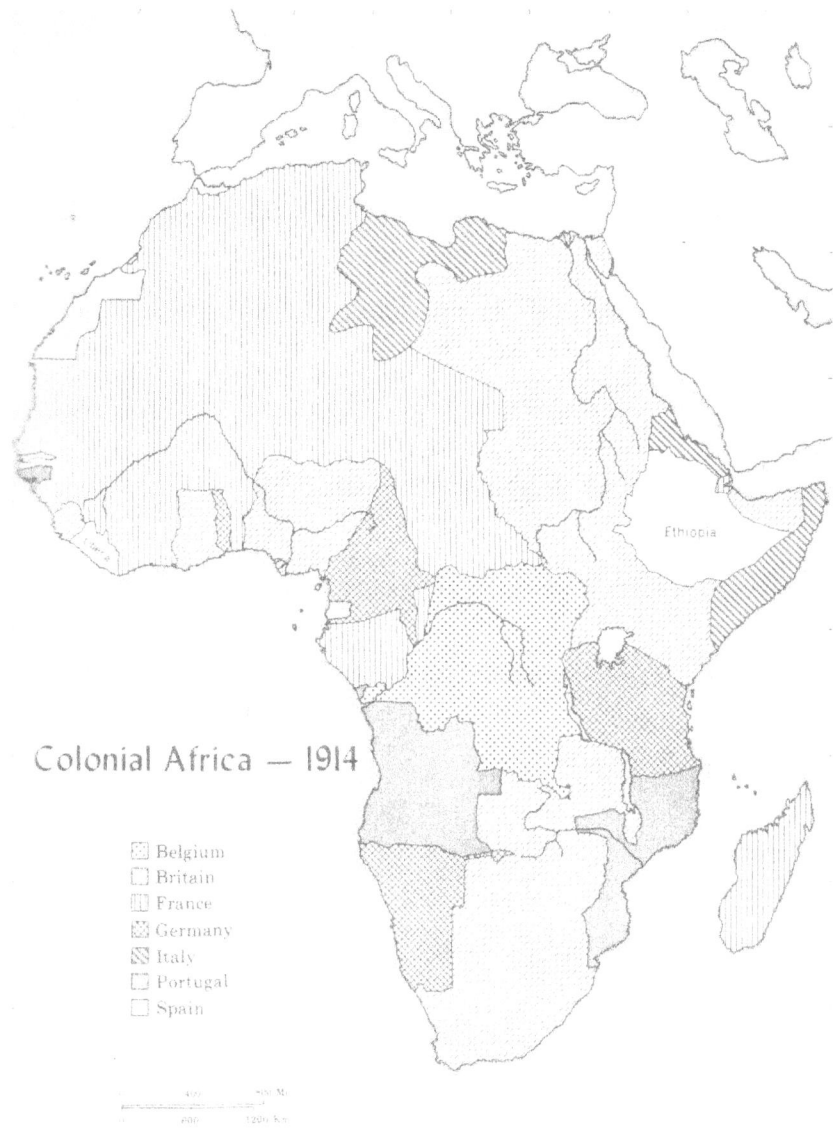

Africa Partitioned after the Berlin Conference of 1884 -1885

****

From *__Africa Since 1800__*, by Roland Oliver and Anthony Atmore:

The world of 1939 had been changed almost beyond recognition. Political, economic and military leadership had passed for the West European countries to the two super-staes, America and Russia. (pg 206)

World War II and European Prestige

Most significant for the subsequent spread of nationalism in Africa was what can be termed the psychological effects of the war. The mental attitudes of Europeans and Africans towards each other were greatly changed by the war. Previously, Europeans had been abke ti dominate Africans, not only because of their more advbanced military and economic techniques, but also becaused they believed that they were superior and invincible. And most Africans believed this too. The Second World War, even more than the First, completely shattered this myth. (pg 208)

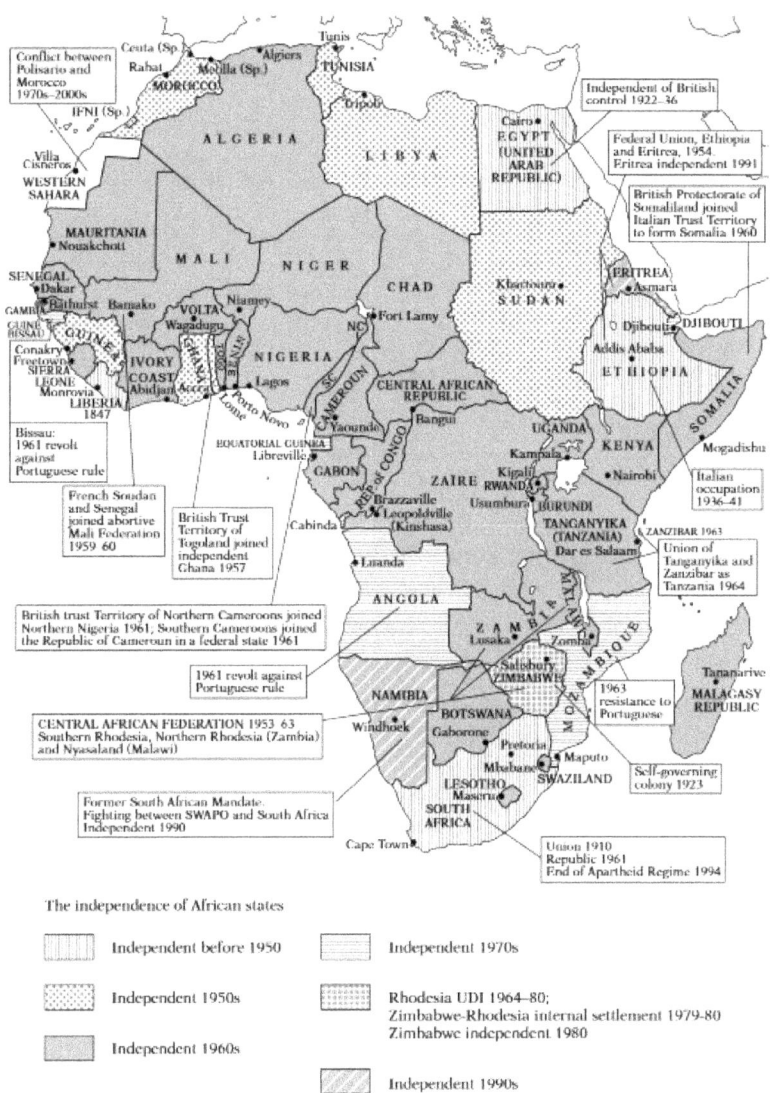

Independence in Africa

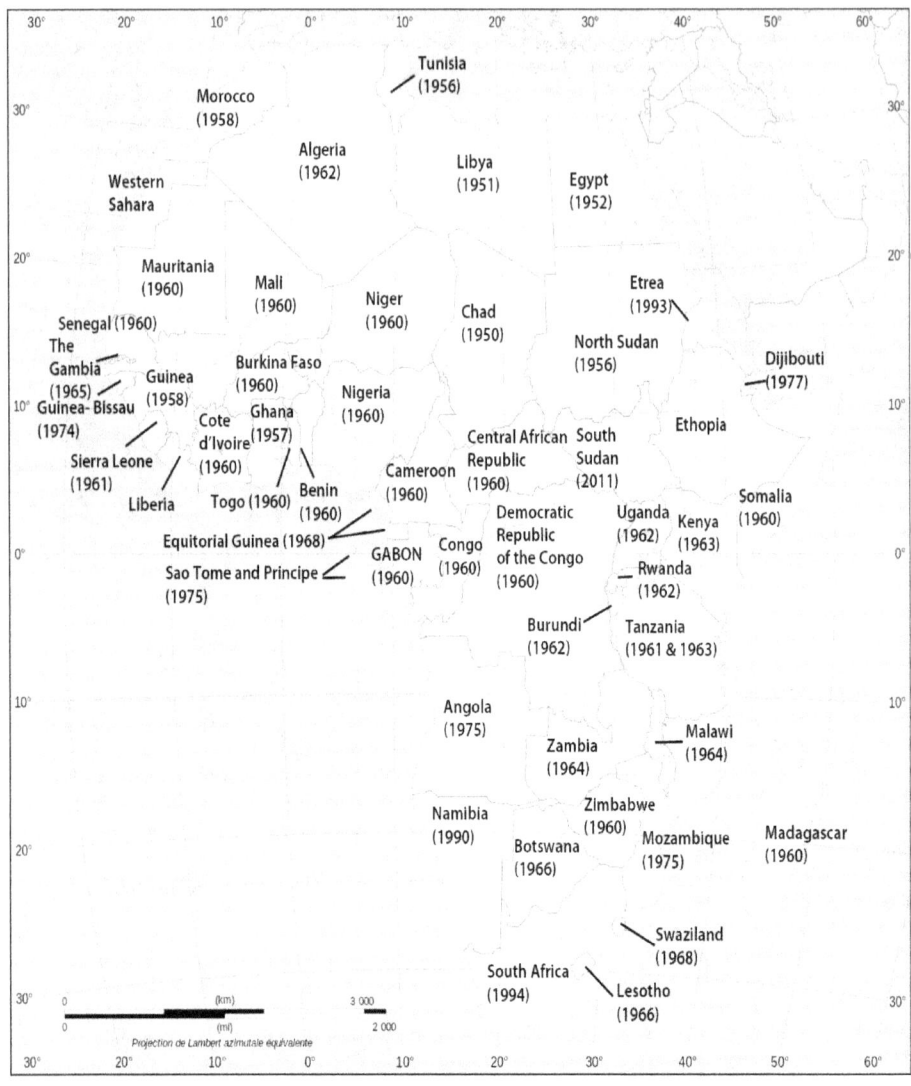

Independence in Africa

Africa, its countries' natural resources and African Scientific Institute (ASI) Fellows

	Country	*ASI Fellows*	*Resources*
1	Algeria	x	Oil, natural gas, iron, phosphates, lead, zinc, silver, copper, gold
2	Angola	x	Petroleum, diamonds, iron ore, phosphates, copper, feldspar, gold, bauxite, uranium
3	Benin	x	Not large quantities of iron ore, phosphates, chromium, rutile, clay, marble, diamonds, and limestone
4	Botswana	x	Numerous minerals are known to occur in significant quantities, but with the exception of diamonds, copper and nickel, they are mostly untapped. These are: salt, soda ash, potash, coal, iron ore, manganese and silver. Botswana has the largest known coal reserves in Africa, although they are of low quality.
5	Burkina Faso	x	manganese and gold
6	Burundi	x	Nickel, uranium, rare earth oxides, peat, cobalt, copper, platinum, vanadium, niobium, tantalum, gold, tin, tungsten, kaolin, limestone.

7	Cameroon	x		oil, bauxite, iron ore
8	Cape Verde	x		salt, pozzolana (sandy, volcanic ash used to produce cement) and limestone
9	Central Africa Republic			diamonds, uranium, timber, gold, **oil**
10	Chad			oil, uranium, natron, kaolin, gold, limestone, sand and gravel, salt
11	Comoros			negligible natural resources
12	Congo, DRC	x		34% of the world's coltan reserve and 10% of copper; Silver, Manganese, Cobalt, Gold, Cassiterite, Diamonds, Oil, Zinc, Cadmium, Palladium, Coal, Germanium, Platinum, Wolfamite, Iron, Bauxite, Limestone, Beryl, Phosphates, Natural gas
13	Congo, RDC	x		ranks fifth or sixth, behind Nigeria, Angola, Sudan, Equatorial Guinea and possibly also Gabon; lead, zinc, copper, uranium, magnesium and gold
14	Cote d'Ivoire	x		oil, natural gas, diamonds, manganese, iron, cobalt, bauxite, copper
15	Djibouti			Salt, Perlite, Gypsum, Limestone, Gold, Granite, Pumice, Oil, Diatomite
16	Egypt	x		Oil, natural gas, phosphates, gold and iron ore
17	Equatorial Guinea	x		oil, and small unexploited deposits of gold, manganese, and uranium
18	Eritrea	x		gold, potash, zinc, copper, salt, possibly oil and natural gas
19	Ethiopia	x		Potash, salt, gold, copper, platinum, natural gas (unexploited)

20	Gabon	x	oil, natural gas, diamond, niobium, manganese, uranium, gold, timber, iron ore
21	Gambia	x	titanium (rutile and ilmenite), tin, zircon
22	Ghana	x	gold, industrial diamonds, bauxite, manganese, rubber, oil, silver, salt, limestone
23	Guinea	x	possesses over 25 billion metric tons (MT) of bauxite--and perhaps up to one half of the world's readily exploitable reserves. In addition, Guinea's mineral wealth includes more than 4 billion tons of high-grade iron ore, significant diamond and gold deposits, and undetermined quantities of uranium.
24	Guinea-Bissau		phosphates, bauxite, clay, granite, limestone, unexploited deposits of oil
25	Kenya	x	limestone, soda ash, salt, gemstones, fluorspar, zinc, diatomite, gypsum
26	Lesotho	x	diamonds, building stone
27	Liberia	x	iron ore, diamonds, gold
28	Libya		oil, natural gas, gypsum
29	Madagascar	x	graphite, chromite, coal, bauxite, rare earth elements, salt, quartz, tar sands, semiprecious stones, mica
30	Malawi	x	limestone, unexploited deposits of uranium, coal, and bauxite
31	Mali	x	gold, phosphates, kaolin, salt, limestone, uranium, gypsum, granite **note:** bauxite, iron ore, manganese, tin, and copper deposits are known but not exploited

32	Mauritania		iron ore, gypsum, copper, phosphate, diamonds, gold
33	Mauritius	x	N.A.
34	Morocco	x	phosphates, iron ore, manganese, lead, zinc, salt
35	Mozambique	x	coal, titanium, natural gas, tantalum, graphite, gemstones, bauxite, and gold
36	Namibia	x	diamonds, copper, uranium, gold, silver, lead, tin, lithium, cadmium, tungsten, zinc, salt, hydropower, fish **note:** suspected deposits of **oil**, coal, and iron ore
37	Niger	x	uranium, coal, iron ore, tin, phosphates, gold, molybdenum, gypsum, salt, oil
38	Nigeria	x	the 12th largest producer of oil in the world and the 8th largest exporter, and has the 10th largest proven reserves. It has underexploited mineral resources which include natural gas, coal, bauxite, tantalite, gold, tin, iron ore, limestone, niobium, lead and zinc
39	Rwanda	x	natural resources are limited: gold, cassiterite (tin ore), wolframite (tungsten ore), methane
40	Sao Tome and Principe	x	Sao Tome and Nigeria reached agreement on joint exploration for oil in waters claimed by the two countries.
41	Senegal	x	phosphates, marble, basalt, sandstone, limestone, and iron ore
42	Seychelles	x	N.A.
43	Sierra Leone	x	Diamonds, titanium ore (rutile), bauxite, gold, iron ore, ilmenorutile, platinum, chromite, manganese,

			cassiterite, molybdenite
44	Somalia		uranium and largely unexploited reserves of iron ore, tin, gypsum, bauxite, copper, salt, natural gas, likely oil reserves
45	South Africa	x	gold, chromium, antimony, coal, iron ore, manganese, nickel, phosphates, tin, rare earth elements, uranium, gem diamonds, platinum, copper, vanadium, salt, natural gas. The world's biggest producer of platinum, and one of the leading producers of gold, diamonds, base metals and coal.
46	Sudan	x	oil; small reserves of iron ore, copper, chromium ore, zinc, tungsten, mica, silver, gold
47	Swaziland	x	asbestos, coal, clay, cassiterite, small gold and diamond deposits, quarry stone, and talc
48	Tanzania	x	tin, phosphates, iron ore, coal, diamonds, gemstones, gold, natural gas, nickel
49	Togo	x	phosphates, limestone, marble
50	Tunisia	x	oil, phosphates, iron ore, lead, zinc, salt
51	Uganda	x	copper, cobalt, limestone, salt, gold
52	Zambia	x	copper, cobalt, zinc, lead, coal, emeralds, gold, silver, uranium
53	Zimbabwe	x	Deposits of more than 40 minerals including diamonds, ferrochrome, gold, silver, platinum, copper, asbestos, nickel, graphite, coal, lithium, palladium, vermiculite

****

Do You Know These African Scientists?

Thierry Zomahoun argues that Africa is the birthplace of mathematical sciences, and that this is the continent's single biggest contribution to humanity. *Photo: Sibusiso Biyela*

By Sibusiso Biyela

Wilfred Ndifon of Cameroon solved a 70-year-old immunology conundrum. Bernie Fanaroff of South Africa established the Fanaroff-Riley classification of radio galaxies and quasars. And Noble Banadda of Uganda uses mathematical models to predict what will happen during the many disease outbreaks in Africa.

Chances are you've never heard of these scientists, even though they are among **the most famous minds** on their continent. Thierry Zomahoun, who leads the *African Institute for Mathematical Sciences (AIMS)*, says that the reason you don't know about them is a scar left behind by the whip of colonialism over Africa.

"I would like you, as science journalists, to help the world discover the unique contribution that the African continent has made."

In a plenary talk given on 27 October to a room full of seasoned science writers at the World Conference of Science Journalists 2017, Zomahoun boldly challenged what he says are **long-held beliefs** of the mainstream media that Africa has not been instrumental in building the body of knowledge in mathematics and science.

"Looking at conventional history, one is left with the sense that Africa has made no meaningful contributions to science," he said. "I would like you, as science journalists, to help the world discover **the**

unique contribution that the African continent has made and continues to make to science and human progress."

The voices of Africa

The history of Africa has long been told through the voices of those who colonised it. A new crop of academics is taking back this narrative history by placing the continent at the centre of human progress. The late Ivan Van Sertima, associate professor at Rutgers University, once poignantly wrote that "the nerve of the world has been deadened for centuries to **the vibrations of African genius**."

For instance, consider the Lebombo bone, a well known 43,000-year-old tool used in specialised rituals discovered in Swaziland in 1973. Scientists say that in addition to its religious importance, it is also evidence of the use of complex mathematics in Africa. This piece of history is the basis for Zomahoun's claim that Africa is **the cradle of mathematics**, countering the idea that humans did not develop the capacity for science until they migrated out of Africa.

Zomahoun then listed off a few African scientists, many of whom changed their fields and have been lauded in their countries. When he asked how many in the audience knew them, he was met with **dead silence**—perhaps reflecting the media coverage of the researchers he listed.

Why should you care about African science

Africa is home to two-thirds of the *Square Kilometre Array* radio telescope, which will trace a new network of smaller radio telescopes across the continent to unravel secrets of the early universe. Africa is also host to programmes such as the Maternal and Infant Health Care Strategies Unit in South Africa that looks to reduce maternal deaths, a scourge that affects even developed countries such as the United States.

The reason we don't hear more about these projects, said Zomahoun, is **a Western bias** left by colonialism. His African Institute of Mathematical Sciences is "enabling Africa's talented students to become innovators driving the continent's scientific, educational and economic self-sufficiency."

Organizers of the *Next Einstein Forum*, one of the institute's initiatives, believe Africa's contributions to the global scientific community are critical for progress. Zomahoun touts this programme as the best way to combat negative ideas about African science by bringing together the continent's brightest young scientists to change Africa's future in science for the better.

If such programmes succeed, the continent's contributions will become the world's achievements toward greater human progress. Perhaps in the not-too-distant future, Zomahoun predicted, **you will be able to name several African scientists** as their work graces more science publications.

****

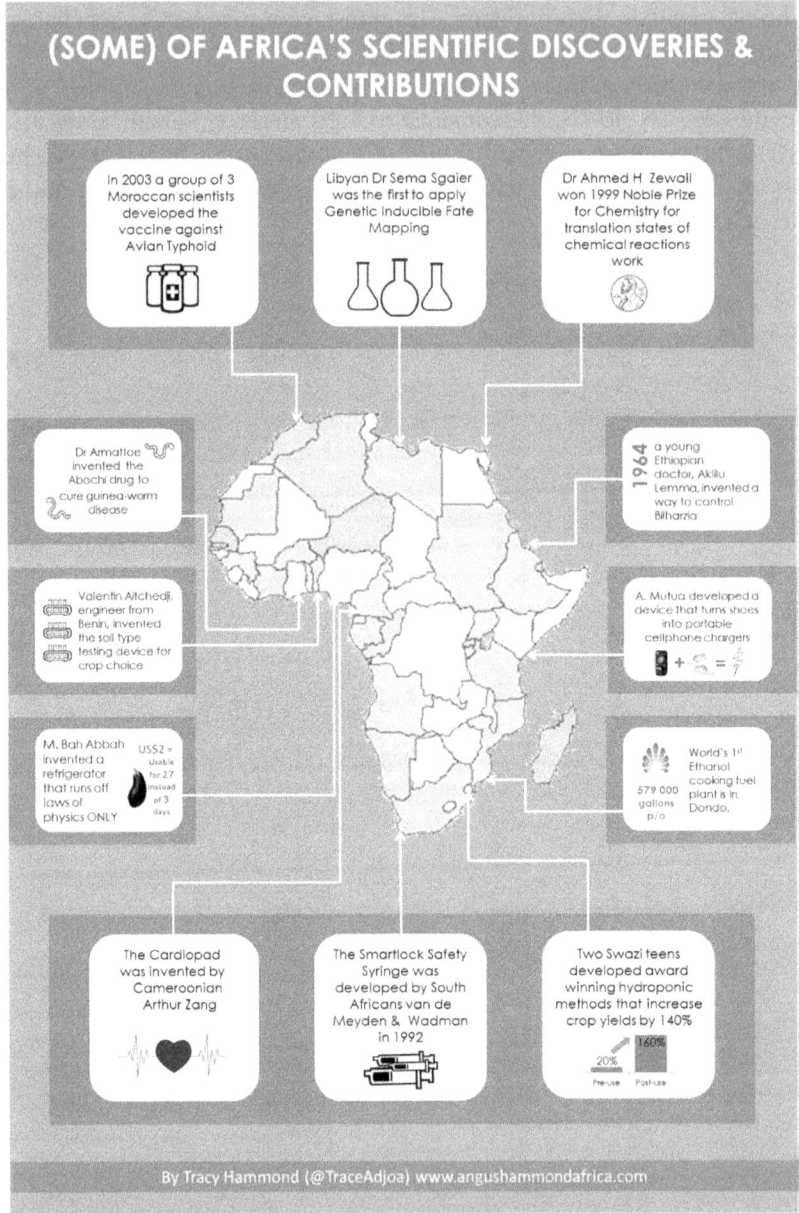

Contributions to research in science and technology by Black women in Africa continues to grow.

A small example of the thousands of Africa's Black people in Science and Technology

Yaye Kene Gassama, Ph.D., *Director General of National Agency of Applied Scientific Research; biotechnology and plant physiology*

Former Minister for Scientific Research, Senegal, Dr. Gassama has also been Chairperson of the African Ministerial Conference on Science and Technology (AMCOST); Vice chair of General Assembly of COMSTECH (Committee on Science and Technological Cooperation); National Coordinator of Biotechnology and Biosafety for Senegal. She has been Professeur titulaire in plant biology department, Faculty of Sciences and techniques University of Dakar, in charge of lectures in biotechnology, microbiology and plant physiology ; responsible of the laboratory of plant biotechnology. Dr. Gassama has also been the National coordinator for UNEP on biotechnology and biosafety and consultant for CORAF-WECARD (Conseil Ouest et Centre Africain pour la recherche et le dévelopment agricoles). She has authored and co-authored 38 scientific publications in international journals.

Justin Ahanhanzo
Physical Oceanography; Coordinator and Team Leader, UNESCO's GOOS-AFRICA Program

Justin is an Oceanographer and Team Leader of the Global Ocean Observing System in Africa (GOOS-Africa coordinator). He is the Technical Secretary of the GOOS Regional Alliances, GOOS-AFRICA Coordinator and Team Leader of the UNESCO Crosscutting Project on the Applications of Remote Sensing for Integrated Management of Ecosystems and Water Resources in Africa.

Shaukat Ali Abdulrazak, Ph.D.
Exec. Sec., National Council for Science and Technology; AgriScience

Prof. Shaukat Ali Abdulrazak is the Executive Secretary of the National Council for Science and Technology (NCST), which is the focal point on Science, Technology and innovation in Kenya. As the Executive Secretary to the NCST, Prof. Abdulrazak is the technical and administrative head of the NCST and is responsible for its day to day management of financial and physical resources as well as the management of human resources and programmes. He also guides the Council in the development and advocacy of Science and Technology policy popularization and promotion of science and technology as a tool for national development. Prior to joining the NCST, Prof. Abdulrazak was the Deputy Vice-Chancellor of the Research and Extension unit of the Egerton University in Njoro, Kenya, a position he held from 2002 – 2007. He also worked as a Senior Lecturer and Associate Professor at this university from 1997-2001. Since 2005, he worked as Professor of Animal Science at Egerton University.

Iman Abuel Maaly Abdelrahman, Ph.D.
University Department Head of Electrical and Electronic Engineering

Dr. Abdelrahman is the Head of the Department of Electrical and Electronic Engineering at the University of Khartoum. She has over sixteen years of experience in teaching and research, and eleven years in operation and maintenance of telecommunication systems. She is the founding CEO of the Sudanese Research and Education Network (SudREN). Dr. Abdelrahman is the CEO of the Sudanese Universities Information Network (SUIN) and the Director of the Information and Communication Technology Department of the Ministry of Higher Education in Sudan. She worked for the University of Khartoum (1999- 2006) as Assistant Professor. In the University of Khartoum she worked as Member of the University Data Network Committee and a member of the Computer Science Advanced Library Committee. She also submitted the proposal and managed the project of the Sudan Engineering Virtual Library which is funded by UNESCO and the University of Khartoum in 2004. She also worked as telecommunication engineer in the Operation and Maintenance Department of Sudan Telecommunication Public Corporation (named Sudatel after privatization) from (1982 -1994), and as a First Planner in the Planning Department of Sudatel in 1994.

Sam Olatunji Ale, Ph.D., *Mathematics*

Prof. Ale has been re appointed for a 2nd term of five years as the Director General and Chief Executive of the National Mathematical Centre, Abuja. He has been commended by the Minister of Education for contributing immensely to the transformation of the Mathematical Science and Mathematical Sciences Education in Nigeria and other areas of human endeavour; for introducing new methods for demystifying mathematics as well as making mathematics simple, real and scintillating to pupils and students and for promoting the image of Nigeria in Mathematical Sciences internationally, especially in the scientific activities in the International Mathematics and Sciences Olympiads. Prof. Ale has produced over 99 publications in learned journals both local and foreign, mostly in Pure Mathematics and Mathematics Education. He has authored 8 published text-books and co-authored 14 others with MAN. He has tutored to graduation over 7 Ph.D's and 15 M. Sc. students and examined the Thesis Defense of over 11 Ph. D. and 150 M.Sc. graduates.

Osama Awadelkarim, Ph.D., *Engineering Science and Mechanics, Nanotechnology*

Dr. Awadelkarim is Professor of Engineering Science and Mechanics and the Associate Director for the Center of Nanotechnology Education and Utilization at The Pennsylvania State University. He earned his B.S. in Physics from the University of Khartoum, Sudan, and his Ph.D. from Reading University in the UK. His research interests are in electronic materials, nano/microelectronics, and nano/microelectromechanical systems. Dr. Awadelkarim has authored/coauthored nearly 200 journal articles, book chapters, books, and conference proceedings. He is a recipient of Shell and the University Prizes from the Sudan, and Fellowships from the International Seminars in Physics and Chemistry (Sweden) and the International Center for Theoretical Physics (Italy). Dr. Awadelkarim worked in the Office of Public Diplomacy and Public Affairs at the Bureau of African Affairs and the Office of Science and Technology Cooperation in the Bureau of Oceans and International Environmental and Scientific Affairs. His assignment in the two offices was to promote interaction and collaboration between African, Arab, and Moslem scientists and scientists in the U.S. He presented talks and seminars at various scientific meetings and workshops in African and Islamic countries and developed a number of science and technology agreements between U.S. government agencies and their African and Islamic counterparts.

African Diaspora

In relative modern times, Black people throughout the African Diaspora have contributed to the development of mankind. Per Keith Holmes book *Black Inventors* we have contributed to advancements in science and technology:

The list of inventors and inventions goes on and on. Raymond Webster (an ASI Fellow) of the United States has documented over 30,000 patents by African Americans. Patricia Sluby (an ASI Fellow) is a certified patent reviewer who worked for the U.S. Patent Office; she has documented hundreds of patents by African Americans.

****

While Katherine Johnson was the featured figure in the movie *"Hidden Figures"*, **there were several Black super-stars of NASA during the 1960s.**

In June 1953, **Katherine Johnson** was contracted as a research mathematician at the Langley Research Center. At NASA, Katherine Johnson started work in the all-male Flight Mechanics Branch and later moved to the Spacecraft Controls Branch. She calculated the trajectory for the space flight of Alan Shepard, the first American in space, in 1959 and the launch window for his 1961 Mercury mission. She plotted backup navigational charts for astronauts in case of electronic failures. In 1962, when NASA used computers for the first time to calculate John Glenn's orbit around Earth, officials called on her to verify the computer's numbers. Ms. Johnson later worked directly with real computers. Her ability and reputation for accuracy helped to establish confidence in the new technology. She calculated the trajectory for the 1969 Apollo 11 flight to the Moon. Later in her career, she worked on the Space Shuttle program, the Earth Resources Satellite, and on plans for a mission to Mars. Katherine Johnson's social impact as a pioneer in space science and computing may be seen both from the honors she has received and the number of times her story is presented as a role model. Since 1979 (before she retired from NASA), Katherine Johnson's biography has had an honored place in lists of African-Americans in Science and Technology.

NASA research mathematician Katherine Johnson at her desk at Langley Research Center. Katherine Johnson, ASI Fellow (a program of the African Scientific Institute) and, pioneering NASA Mathematician, born 1918, died on February 24, 2020 at 101. She was the featured figure in the movie *"Hidden Figures"*.

****

"We Could Not Fail", The First African Americans in the Space Program, by Richard Paul and Steven Moss:

Over three decades, he recruited hundreds of African Americans into the space program. **Clyde Foster** processes telemetry at the Marshall Space Flight Center in 1965 in a photo that appeared in Ebony magazine. As a NASA employee, Foster was a leader in getting jobs and advancing engineering education for African Americans.

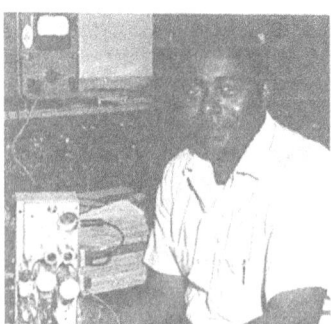

Julius Montgomery, a pioneering African American in the space program. He was the first African American to work as a technical professional at Cape Canaveral He began his career as a member of the "Range Rats," technicians who repaired malfunctioning ballistic missiles.

Dr. Frank A. Crossley, the first African-American earning a Ph.D. in metallurgical engineering. **(1925 – 2018)** Dr. Frank Crossley was a pioneer in the field of titanium metallurgy. He began his work in metals at Illinois Institute of Technology in Chicago after receiving his graduate degrees in metallurgical engineering. In the 1950s, few African Americans were visible in the engineering fields, but Dr. Crossley excelled in his field. He received six patents, five in titanium base alloys. His research greatly improved the aircraft and aerospace industry. He has published over fifty papers in journals and symposia. He is listed in several who' who including Who's Who in America, 41st, 1980-81 through 61st, 2007 Editions.

Morgan Watson (then and now), one of the young black engineers from Southern University. Who helped fill out 'road map' to the moon. He started working at NASA's Marshall Space Flight Center in 1964. Morgan Watson is founder and president of Minority Engineers of Louisiana in Baton Rouge, Louisiana

****

Census Classification Systems Used to Identify Black people

One of the most profound findings in researching information about people of color and their inventions was the classification system devised to accurately document information about Black people. This system was used as a decoy to keep black people from identifying one

another as members of the same race. It was common practice to classify Black people under a host of different categories: Aborigine, African, Afro, Black, Caribbean, Colored, Dravidian, Khaffir, Moors, Mulatto, Native African, Negro, Octoroon, Quadroon, Slave, South African, West Indian, ...etc.[34]

Today it is estimated that there are about 40,000,000 Black people in the USA. There are about 25,000,000 in Central America and the Caribbean; 150,000,000 in South America; 8,000,000 in Europe; 20,000,000 in China; 30,000,000 in Indonesia; one billion in Africa.

The African Diaspora represents a voluntary and involuntary migration of Black people throughout the world. Whether this migration occurred through historical slave trades or modern day voluntary movement to find better education and employment opportunities, Black people seek to form partnerships to advance countries within Africa.

Ascertaining the level of quality among a skilled workforce is critical to the development of African countries. To move forward among developing countries, institutions, organizations and corporations must develop human resources to keep up with emerging technologies. These individuals are critical for job sustainability and growth, food security, health care delivery, and infrastructure development. Skilled personnel are also vital to train the next generation of people who will be responsible to sustain the health and well-being of African countries.

Countries throughout the world need skilled workers to sustain their vitality. The countries that can best afford skilled workers often drain skilled workers from African countries. Not only do African countries need support from its past citizens who were forced from Africa as slaves and are now highly sought after skilled workers, Africa countries need support from current skilled Africans who voluntarily leave Africa to seek better paying jobs. These individuals also migrate overseas because of better education opportunities to stay abreast of the most advance to live in other more politically stable countries. Often, graduate students and highly skilled professionals find they do not possess the latest laboratory equipment to perform tests and research studies. Beyond lip service, many African countries have not yet fully committed themselves to growth through the better use of developments in science and technology.

African countries (also in most countries throughout the world) are not only concerned about international migration of its highly educated and skilled workforce; these countries also have concerns about migration of these people from rural areas to urban areas to seek better opportunities, leaving rural areas in dire need of support of skilled workers.

There are bright sides to the migration of Black people internationally. Several countries which have large populations of Black people find themselves assisting in the development of Africa. Recently, Senegalese President Wade opened the next FESMAN III (now "World Black Arts and Cultures Festival") event to be held in Senegal and four other adjoining countries during December 2010. At first during the 2009 FESMAN III, the planners of this massive event anticipated 80,000 people throughout the African Diaspora to attend and participate in the festivities. So countries like China and some European countries decided to pitch in to make sure attendees would be properly housed and events would have stages and exhibition space in which to perform and display their wares. The excitement was so great that President Lulu of Brazil announced that Brazilians would attend and participate. In fact, Brazil would co-sponsor this event. As we all know, President Lulu identifies with Africa, as well as many of Brazilian citizens who are members of the African Diaspora. Then an astounding number of additional worldwide artists and excited on-lookers wanted to attend. Anticipated attendees and participants to this event now grew to 250,000! Senegalese and Brazilian leaders wisely decided to delay presenting this event until December 2010.

There are countless stories of Black people throughout the African Diaspora who reach out to assist Africa, the continent of their fore fathers and mothers.

Experts in science and technology from Brazil are partnering with African countries to enhance developments of projects in Africa[35].

Cuba also identifies with Africa[36], as most of its citizens are part of the African Diaspora. Cuban doctors practice in African countries today, saving lives and training other indigenous health care workers. Cuban military fought with Namibian and Angolan forces to defeat colonializers to free Namibia and sustain the freedom of Angola.

West African countries such as Senegal offer shelter and even citizenship to Haitians who were devastated by the recent earthquake.

African Americans identify with Africa and lend assistance wherever possible.

The African Scientific Institute (ASI) has moved forward over the past fifty (50+) plus years to engage Black people in the world of science and technology. Our database is global and extensive. It has been used for ASI to successfully engage in activities and projects in the African Diaspora.

Some people of ASI has quietly helped to establish other resources such as the National Society of Black Engineers (NSBE, USA), which now has chapters in Africa and the Northern California Council of Black Professional Engineers (NCCBPE). United States groups like the National Organization of Black Chemists and Chemical Engineers (NOBCChE), National Society of Black Physicists (NSBP) , American Association of Blacks in Energy (AABE), National Medical Association (NMA) are partnering with counterparts in Africa.

ASI believes that Black people throughout the African Diaspora can further assist in the development of African countries. First, we need to assess the quantity and quality of the highly educated and skilled workforce domestically and internationally among Black people within the African Diaspora. We then need to network these individuals among themselves to determine their willingness and availability to assist the development of African countries. To this end, ASI has moved forward through its *ASI Fellows Program.*

The African Scientific Institute (ASI) is helping to join hands among Black brainpower among scientists and technologists throughout the African Diaspora. ASI Fellows Program inspires young people to pursue careers in science and technology, while simultaneously representing a network of scientists, engineers, technologists, and health professionals, mathematicians. Through its *ASI Fellows Program*, we represent a network of 1,500+ "super stars" in science and technology from 57 countries. They have continuously demonstrated their efforts to achieve higher goals in their areas of expertise and serve as a standard of accomplishments. They are role models who encourage our youth to further engage in science and technology. Our youth can identify with these individuals and be encouraged that as they start their journey, they can also achieve high accomplishments, knowing that others have achieved similar goals.

NORTH AMERICA

CANADA

In 2016, close to 1.2 million people in Canada reported being Black

Since 1996, the Government of Canada's annual Black History Month campaign encourages Canadians to learn more about the Black experience in Canada This brief overview documents some of the events that helped to shape the contributions of people of African and Caribbean descent to the settlement, growth and development of Canada.

The first person of African heritage to come to what is now Canada arrived some 400 years ago. In 1604, Mathieu Da Costa arrived with the French explorers Pierre Du Gua De Monts and Samuel de Champlain. Da Costa, a multilingual interpreter who spoke English, French, Dutch, Portuguese, and Pidgin Basque, provided an invaluable link with the Mik'maq people encountered by the Europeans.

Throughout the 1800's, a number of historic Black communities were established across Canada. Some of these communities came as a result of war. Also, between 1800 and 1865, approximately 30,000 Black people came to Canada via the Underground Railway – the network of secret routes and safe houses used by enslaved Africans to escape into free American states and Canada with the support of abolitionists and their allies

A small example of the thousands of Canada's Black people in Science and Technology:

E. Oscar Brathwaite
Technical and Vocational Education and Training

Oscar Brathwaite has worked in diverse areas of education, training and consulting for over thirty five years, both in Canada and Southern Africa (Botswana, Swaziland, Zambia and Zimbabwe). He also made presentations at many international conferences in Africa and South America. He has occupied the positions of department head, teacher trainer, lecturer, curriculum planner, developer and mentor. His areas of expertise include program and curriculum design, development, review, delivery and implementation. His expertise also include feasibility studies, institutional strengthening, capacity building, project evaluation, learning materials evaluation, teaching methods and professional development.

Olive James, MSc
Immunology

A retired senior scientist with 35 years of experience in immunology and vaccine research and development. Her experience includes leading a team within one of the largest global vaccine manufacturers, where she was responsible for a group of highly trained bench scientists in the areas of assay development including antibody-based and cellular-based assays that were used in all phases of research and development. As a subject matter expert Ms. James currently provides Bioanalytical guidance, to support product development in growth-stage companies. Ms. James specialties: Analytical Method Development • Characterization of monoclonal and polyclonal antibodies • Quantification of antigens as active ingredients and as contaminants of biological products • Quantification and analysis of animal sera for pre-clinical and clinical evaluation • Assay development and evaluation of animal models for Non Clinical Safety testing • Development of potency tests (in vivo and in vitro) • Development of biological assays for clinical evaluation of human sera. Her post retirement activities also include mentorship of young scientists.

UNITED STATES

US 2010 Census: 44 Million Black people in USA--

The African Scientific Institute (ASI) believes it can support efforts to reverse the brain drain of African countries through visions of public-private partnerships being formed to engage in on-the-ground projects that integrate academic institutions with government agencies and private corporations and developers. As an example, when health care delivery systems are enhanced, academia has better places to train students and employ professional health care workers. When manufacturing is needed to produce containers for products, engineering students have places to train and professional engineers have places to work. When transportation systems are being developed, students have places to train and professional urban planners and engineers have places to work. As African countries develop in ICT, Space Science, Alternative Energy source usage, Ecology stewardship…there will be opportunities for students to learn and highly skilled people to work. Black people throughout the African Diaspora will come together when efforts to develop in African countries are being actualized.

Through networking together, we can see what each other is doing to help African countries to move forward. ASI believes that development projects should be appropriate for the future of African countries. These projects should be sustainable and environmentally sensitive. One such project in which ASI is engaged involves working with UNESCO's GOOS-Africa on such efforts to realize partnering efforts between ASI and African countries to assist in using satellite imagery to monitor impacts to African coastlines due to climate changes.

Black people in Africa and the African Diaspora can partner to assure educational standards are met for primary and secondary education, as well as centers of excellence occur within African academic institutions of vocational training, colleges and universities. On-the-ground projects should include hands-on experiences for students of these academic institutions.

As Black people throughout the African Diaspora join hands with Black people in Africa, working on projects and activities, bonding between Black people solidifies. Synergistic growth of ideas and visions occur about what can further be accomplished to create excitement about Africa.

A small example of the thousands of USA's Black people in Science and Technology:

Julian Earls, Ph.D.
Ret'd Center Dir., NASA Glenn Research Center, Physicist

Dr. Earls retired in 2006 as Director of the National Aeronautics and Space Administration's Glenn Research Center at Lewis Field in Cleveland, Ohio from 2003 to 2005. Glenn Center is engaged in research, technology and systems development programs in aeronautical propulsion, space propulsion, space power, space communications, and microgravity sciences in combustion and fluid physics. While managing an annual budget of approximately $773 million, he oversaw a workforce of 1,920 civil service employees and 1,300 on-site support service contractors. The center consists of 24 major facilities and over 500 specialized research facilities. Dr. Earls has been a noted Health Physicist and Radiation Specialist. He authored the first Health Physics Guides within his institution; served as a member of the Launch Team for Apollo XIII Lunar Program; served as the first African American Radiation Specialist in the New York Regional Office of the U.S. Atomic Energy Commission (now the U.S. Nuclear Regulatory Commission). He is co-founder of an organization whose members make personal contributions to raise $1 million for scholarships to black students who attend black colleges. Dr. Earls has been inducted into National Black Colleges Hall of Fame, along with Justice Thurgood Marshall and Dr. Martin Luther King, Jr.

Rodney C. Adkins, Ph.D.
Information Technology and Services; Electrical Engineering

Dr. Adkins is President of 3RAM Group LLC. Formerly, he was Senior Vice President of IBM, from 2007 until 2014. He was the company's first African American Sen. VP and corporate officer. In his more than 33-year career, Dr. Adkins has held Sen. VP roles, including Strategic Partnerships and Corporate Strategy from 2013 to 2014, responsible for leading transformation across IBM and developing strategies for a new era of computing, new markets and new clients. Since 2009, Dr. Adkins was previously Sen. VP of Systems and Technology Group (STG), a $20B business segment, responsible for all aspects of IBM's semiconductor, server, storage, and system software businesses, as well as the company's Integrated Supply Chain and Global Business Partners organizations. He was a leading innovator in solutions ranging from mobile devices to the world's largest supercomputers. His major contributions included helping to develop the Personal Computer industry, launching the IBM ThinkPad mobile PC, leading IBM's POWER business to become the dominant player in the UNIX market, and helping to pioneer what became IBM's portfolio of Smarter Planet solutions. In 2005, Dr. Adkins was inducted into the U.S. National Academy of Engineering (NAE). He continues his national focus on STEM as the Chair for the National Science Foundation's committee on STEM workforce strategy and as the Co-Chair for the NAE committee on Engineering Education Workforce Continuum.

George E. Alcorn, Ph.D., *Physics*

Dr George Alcorn is an Asst. Director at NASA's OSSMA. At NASA he has served as Deputy Proj. Mgr. for Space Station Adv. Dev., Chief of the Technology Commercialization Office, Proj. Mgr. for the Airborne Lidar Topological Mapping System (ALTMS), Mgr. of the In Step Flight Experiment Program, Deputy Mgr. for Center's Code R research program, and Deputy Div. Chief for NASA's Technology Integration Div.. He led the program that established NASA 's first technology incubator in Baltimore. He earned his BS degree in Physics with honors while earning eight letters in basketball and football. He earned his MS in Nuclear Physics in 1963 from Howard Univ., after nine months of study. In 1967, he earned a Ph.D. in Atomic and Molecular Physics. Dr. Alcorn has earned over 20 patents. For over 25 years, he served as an adjunct full professor, teaching physics and electrical engineering. He is a pioneer in using particle tech-niques like plasmas, ion beams, and sputtering methods to make high performance semiconductor devices. He invented a radical x-ray spectrometer using thermomigration of aluminum. Among Dr. Alcorn's numerous awards is a NASA medal for his work in recruiting minority scientists and engineers and helping small businesses have successful research programs. For 17 years he spent his Saturdays teaching courses like C++ to bright inner

Donna Auguste,
Software Engineering, Artificial Intelligence

Co-founder of Freshwater Software, Inc. (acquired for $147 million in 2001), is an extraordinary computer scientist. In 1984, she worked for Intellicorp, a Silicon Valley start-up that was part of the rush to commercialize "expert systems", artificial intelligence programs that were emerging from research labs. Ms. Auguste then joined Apple computer, where she managed the elite team of programmers who helped create the industry's first PDA, Apple's pen-based/hand-held computer called the Newton. In 1996 Ms. Auguste co-founded Freshwater Software in Boulder, Colorado. Her team developed a product called SiteScope which is used by systems administrators who manage Internet and Intranet Web servers. Ms. Auguste is also a devoted Christian and musician. She plays bass guitar and piano. After Freshwater was acquired, Ms Auguste devoted more of her time to her non-profit Global Outreach organization, Leave a Little Room Foundation (www.LeaveaLittleRoom.org), installing solar electricity and building schools in African villages, and building hospitals and houses in Mexico. She does research in robotics and renewable energy resources, and plans to publish her first novel in 2008. Ms. Auguste is a graduate of the University of California at Berkeley, with a B.S. in Electrical Engineering and Computer Science.

Gaurdia Banister, R.N., Ph.D., *Executive Director and Specialist in Psychiatric Nursing*

Dr. Banister is the first Exec. Dir. of the newly established Institute for Patient Care at the Massachusetts General Hospital in Boston. She is responsible for advancing the Institute's new vision for interdisciplinary education and research which is centered on a commitment to meeting patients' needs and advancing nursing and the health professions. Prior to this position, Dr. Banister was the Senior VP for Patient Care Services and Chief Nursing Officer at Providence Hospital in Washington, D.C. She is the principle investigator for a program funded by the Health Resources Services Administration that provides a comprehensive array of services to support disadvantaged high and college students who aspire to be registered nurses. Dr. Banister also been involved with District of Columbia's Nurses Association to enact legislation to ensure that nurses with emotional and/or substance abuse problems receive adequate care and rehabilitation. She also was the chairperson of the Committee on Impaired Nurses that provides a safety net for nurses with psychiatric and substance abuse problems. In 2006, Dr. Banister was selected as Johnson and Johnson Wharton Nurse Fellow. In 2001, she was selected as a Robert Wood Johnson Executive Nurse Fellow. She also received the District of Columbia Nurses Association Practice Award in 1998.

Robert "Pete" Bragg, Ph.D., *Univ. Prof. Emeritus, Material Scientist, Physicist*

(- 2017) When someone says"carbon", what do you think of? Coal, right? Or maybe diamonds. But when the nose cone of a space shuttle withstands heat and impact upon reentry into the earth's atmosphere, that's carbon, too. When you enjoy the resilience of a graphite tennis racquet or the purity of water filtered by some new system, you may gain insight into the applications of Dr.Robert H. Bragg's forty-plus years of research on the structure and physical properties of carbon. In his research in x-ray diffraction and small angle x-ray scattering techniques, he collaborated with scientists in several laboratories in the U.S., as well as in Japan, France, Germany, Algeria, and Nigeria. Prior to joining academia, Dr. Bragg was a manager of research in metallurgy at the Research and Development Laboratory of the Lockheed Missiles and Space Company, Palo Alto, California. He was responsible for research activity in solidification, metallurgical composites, mechanical and thermo-physical properties of refractory compounds, phase diagrams determinations, as well as research in the metallographic, x-ray diffraction, and electron microscope laboratories. Dr. Bragg was a Fullbright Scholar at Olafemi Awolowo University, Ife, Nigeria (1993-4). He also worked with the Collaborative Access Team at the Advanced Photon Source, Argonne National Lab (Fall, 2000).

George R. Carruthers, Ph.D.
Astrophysics

Dr. Carruthers held a position of Rocket Astronomy Research Physicist from 1964 to 1982. An inventor as well as physicist, he was instrumental in the design of lunar surface ultraviolet cameras, using the color spectrum of substances to detect their constituent parts. In 1969, he patented an image converter for detecting far ultraviolet electromagnetic radiation, which was first used in sounding rocket flights, including one in 1970 which made the first detection of molecular hydrogen in interstellar space. He then invented the Far Ultraviolet Camera/Spectrograph, a device which would examine both the Earth's upper atmosphere and deep space from a location that would avoid the distortions created by Earth absorption of ultraviolet radiation. By 1972, Dr. Carruthers's camera/spectrograph was constructed. Cdr. John W. Young carried the device aboard the Apollo 16 mission and placed it on the surface of the moon. Over 200 pictures of the Earth's atmosphere and geocorona, as well as of the Milky Way and deep space, were taken from this observatory. He also invented photometry ultraviolet cameras and spectrographs for rockets and satellites, several electronic imaging devices and other detectors for space astronomy and upper atmosphere physics research. In 1986, one two of Carruthers' inventions instruments captured an ultraviolet images and spectra of Halley's Comet.

Julius W. Garvey, M.D.,
Thoracic, Cardiothoracic Vascular Surgery; son of Marcus Garvey

Dr. Garvey continues to combine his busy surgical practice (Thoracic, Cardiothoracic Vascular) with community service. He is the youngest son of Marcus Garvey. Dr. Julius Garvey is a founding member, and is currently Chairman of the Marcus Garvey Committee International, Inc., an organization that serves to improve the economic, cultural, educational, and spiritual condition of Africans all over the world. Additionally, he serves on the Boards of various other organizations such as, The Board for the Education of People of African Ancestry, The Zumbi Foundation, and the Brotherhood. Over the years, he has worked in conjunction with the various Ministries, the Department of Corrections, and the University of the West Indies, on issues concerning the education of the Jamaican youth, the building of schools, the transfer of books and medical supplies, and a medical student exchange program. He lectures on African History and culture and on the legacy of Marcus Garvey, in the United States, Canada, and the Caribbean. His audience includes Junior High, High School, and college students, as well as, social communities, religious, political and national organizations.

Thomas (T.C.) Charles Adams, III
Energy conservation and technology development

As Director, Energy Division North Carolina Department of Commerce, T.C. Adams led and directed the activities of the Energy Division, the state's primary agency for energy policy and administration, in its statewide program for energy conservation and technology development. He Created and Chaired the state Interagency Alternative Fueled Vehicle Task Force, resulting in the purchase of over 1,200 ethanol, propane, natural gas and electric vehicles for the state fleet, development of 6 state ethanol fueling stations, an ethanol supplier network, an electric charging-station and several natural gas stations. He also served for 4 years as Board member, State Energy Advisory Board, United States Department of Energy, established by U.S. Congress in 1990 to advise the legislative and executive branches on the energy needs of states and effects of USDOE programs and policies on states; member of the subcommittee on National Laboratories. T.C. Adams Created and sponsored the Energy Technology Advisory Group (ETAG) consisting of eight university and research laboratories across North Carolina for pursuing federal/industry grants in energy research and technology development. ETAG was given legislative mandates for reviewing the design and construction plans for $40 billion in new school and university construction.

James E. Allen
Ret'd Mgr., Lab Space Flight Operations Facility.

(- 2006) Former manager of the Jet Propulsion Laboratory's Space Flight Operations Facility (SFOF) wass responsible for providing for the maintenance and operation of the SFOF environmental and utility equipment in support of Space Flight Mission Operations. Professionally, Jim has been employed in almost forty years of progressive engineering and management experience from aeronautical engineering to mechanical engineering; from electrical engineering to computer science; and from marketing to technical management. Over the last thirty five years, Jim has been a pioneer in the Space Exploration program at the Jet Propulsion Laboratory since the late fifties and has worked on the design and development of space craft with missions to the Moon, the Planet and beyond. He was Operations Project Engineer for Worldwide Deep Space Tracking Station in support of the space flight missions to Mercury, Venus, Jupiter and Saturn. He was involved with the Europeans in their Giotto mission to Halley's Comet and the planned Ulysses mission over the poles of the Sun.

 **Cheistopher S. Boxe, Ph.D.**
Environmental Science and Engineering

His interdisciplinary focus is on regional and global biochemical and physical evolution of planets on geological and shorter timescales by modeling, laboratory, and remote sensing techniques. Dr. Boxe serves on the African American Resource Team (AART) at NASA's Jet Propulsion Laboratory, contributing to JPL's Diversity Programs Initiative, which seeks to foster an inclusive environment, where the differences and similarities of individuals are valued and respected, ensuring full utilization of the talents and capabilities a diverse workforce. On a voluntary basis he gives scientific presentations basis to elementary schools and professional organizations about his work. Dr. Boxe works on collaborative practical science projects and field trips with a multitude of select elementary, junior high, and high schools in Los Angeles with the main goal of gearing underrepresented peoples towards the hard sciences (*i.e.*, with a special emphasis on Earth and Space Sciences).

BLACKS IN LATIN AMERICA

Latin America Science: The Long View
Jorge Cañizares-Esguerra and Marcos Cueto, September 25, 2007

Until recently, Latin American science has been dismissed as the underdeveloped and dependent poor cousin of science in the developed world.[1] But Latin American scientists have long made important contributions, and the history of science in Latin America provides a sharp lens for viewing the history of the region's relations with the rest of world.

Before the European conquest of the sixteenth century, indigenous civilizations in the Americas had developed very sophisticated knowledge of their surrounding natural and physical worlds.[2] Knowledge of agriculture, medical botany, astronomy and metallurgy was particularly well developed in the Andes and in what is now Mexico and Central America. Unfortunately, though we know the broad outlines of many of these accomplishments, few of the details have been recovered and even less is known about how these systems of knowledge were preserved and adapted in the wake of European colonization. Indigenous knowledge was lost in part because it had been inextricably woven into indigenous religions. The European conquerors persecuted practitioners of these religions and the learned indigenous elite readily embraced Hispanic acculturation. Most indigenous knowledge had been transmitted orally, and this transmission became precarious when indigenous societies came under colonial control.

MEXICO

2002 Black Population: App. 1,000,000

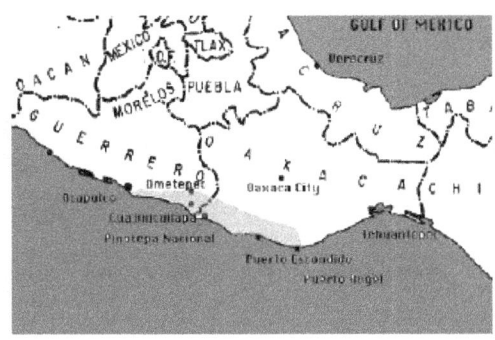
Area of Costa Chica

..the Mexican city of Cibola was founded by a Black man named Esteban el Negro (Steven the Black), a Moor from Spain

...Blacks had important roles in Mexico's military and helped gain its independence from Spain

...the song 'La Bamba' by Los Lobos was originally a song sung by African slaves as they worked in Veracruz

...Bamba is the name of an African tribe in Angola

...Mexico employed more African slaves than any other nation in the western world

...Veracruz, Campeche, Panuco and Acapulco were the main ports for the entrance of African slaves

...Most of the roads, bridges and cathedrals were erected by Black people

...In the 16th century, Afro-Mexicans made up 71% of Mexico's population

...The offspring of African/AmerIndian integration was called jarocho (wild pig), chino or lobo (wolf)

...Vicente Guerrero (El Negro Guerrero), who was Mexico's 2nd President, helped abolish slavery

...Under the dictatorship of Porfirio Diaz Black people were not allowed to immigrate into Mexico

...Many Black communities bear names related to Africa such as Mozambique, Cerro del Congo (Congo Hill) and El Mulato

...Guerrero, Oaxaca, Tabasco and Veracruz are where most Afro-Mexicans live today

...Negro is viewed as derogatory and is no longer used but rather Moreno (Brown) when the subject of Afro-Mexicans is brought up

...Light-skinned Blacks are known as blanquitos (Little Whites) and are the most privileged of Afro-Mexicans

This section focuses on the people of African descent in Latin America. There is confusion as to who should be considered Black or what country should be a Latin American nation. Latinos or Latin Americans includes people from the nations in the western hemisphere whose number one language is the Latin derived Spanish. Brazil, a Portugese speaking country is included because of the similar history with Spanish-speaking nations. English-speaking Belize, although technically not a Latin American country, will be included as well because of its ties to the Central American Garifuna culture. Blackness is not an American thing, its a global thing...
Tatyana Ali (Panama)

CENTRAL AMERICA

Identity of Black People Recognised, But Needs Neglected
By José Adán Silva

MANAGUA, Oct 8, 2010 (IPS) - Although their human rights are increasingly recognised, blacks in Mexico and Central America

are the poorest and most marginalised people in Latin America, according to experts.

"People of African descent in Central America and Mexico are among the most vulnerable, poor and excluded on the continent," Alta Hooker, vice-chancellor of the University of the Autonomous Regions of the Caribbean Coast of Nicaragua (URACCAN), told IPS.

"Ever since the first World Conference against Racism and Discrimination (held in Durban, South Africa in 2001), our people have gained recognition of their human rights, but have not seen much progress in terms of having their social and economic needs met," Hooker said.

Belize, Costa Rica, El Salvador, Guatemala, Honduras, Mexico, Nicaragua and Panama have signed international agreements to protect the living standards of blacks and ensure they are equal to those of the rest of the population, but in practice these rights "are ignored," she said.

In spite of legal and institutional advances towards recognition of the ethnic, cultural and social identity of people of African descent over the last 20 years, their economic status and poverty rates in Central America are worse than in other subregions of Latin America, she said.

Hooker is one of the authors of the study "Derechos de la población afrodescendiente de América Latina: desafíos para su implementación" (Rights of the Afro-descendant Population in Latin America: Challenges to Implementation) that was presented Wednesday at a workshop on the issue in Managua.

The workshop was sponsored by the United Nations Development Programme (UNPD) as part of its regional project, Afro-descendant Population of Latin America.

"A complete legal framework exists, which some countries respect more than others, however. Rights are well established on paper; what is lacking is the will to enforce them," Hooker emphasized.

"There are more than 159 million Afro-descendant people in Latin America, but to date, nine years after the Durban conference, official statistics still cannot tell us how many we are, where we live and what we do. We are the only ones aware of our situation," she complained.

Hooker's views are shared by Dorotea Wilson, general coordinator of the Afro-Latin American and Afro-Caribbean Women's Network (RMAA), a Nicaragua-based organisation with branches in 24 countries of the region.

According to Wilson, inequality is still abysmal, and it is even worse for black women.

"Since Durban, very little has changed for women of African descent in the region. Eighty percent of the more than 150 million Afro-descendants in Latin America continue to live in a state of poverty and have no opportunities to improve their lives, due to ethnic and racial discrimination," she told IPS.

Blacks in Central America "continue to be subjected to forced displacement, illegal migration, the criminalisation of young people and genocide under the guise of fighting crime," she said.

Sidney Francis Martin, head of the Honduras-based Central American Black Organisation (ONECA), expressed a similar view.

"For 20 years we have been signing agreements, treaties, and international conventions on human rights, but in spite of them all, our people grow poorer every day," he told IPS.

"Our identity as Afro-descendants is recognised and respected, and laws are approved on our behalf, but our rights to decent work, better schools and political representation continue to be denied. In these areas we are still invisible," he said.

In the view of Ricardo Changala, a human rights adviser appointed by the U.N. Office of the High Commissioner for Human Rights (OHCHR), the conditions of inequality and poverty endured by blacks "are practically the same throughout the continent.

"Every country has made progress in one way or another, but the gap is still wide," he told IPS.

According to Changala, there have been positive changes in the legal framework in support of the rights of people of African descent, but many challenges remain.

"It's a more than 400-year-old problem that has only begun to be addressed in the last 20 or 30 years, but I am optimistic. I think these barriers will be overcome, so long as the black population continues to claim its rights and make itself heard. The silence has been broken, and that itself is a big step forward." he said.

According to the study presented in Managua, social inequality, poverty and exclusion are among the major problems faced by Afro-descendant populations in Central America and Mexico.

Official statistics cited in the study indicate that 46 percent of the Central American population, some 20 million people, are poor, and one in five of these are extremely poor.

Silvia B. García, the coordinator of UNDP's regional project, said the study set out to measure the gaps between international conventions on human rights and the actual protection and promotion of the rights of black people.

"The study was also used to assess the real and effective compliance with the instruments in signatory countries," García told IPS.

The goal was "to identify the different degrees of visibility of the Afro-descendant population in the countries covered by the study, as well as the diversity or absence of affirmative action, and the varying degrees of progress made by national legislatures in tackling these barriers," she concluded. (END)

Belize

347,000 overall population (Black Population: App. 90,000)

The only English-speaking country in Central America, Belize was a British colony for more than a century and was known as British Honduras until 1973. It became an independent nation in 1981. Belize is a member of the Caribbean Community (CARICOM) and the Sistema de Integración Centroamericana (SICA) and considers itself to be culturally both Caribbean and Central American.

The small, essentially private enterprise economy is based primarily on agriculture, agro-based industry, and merchandising, with tourism and construction assuming greater importance. Sugar, the chief crop, accounts for nearly half of exports, while the banana industry is the country's largest employer. Citrus production has become a major industry along the Hummingbird Highway. More recently, discoveries of petroleum deposits in the Cayo District and possible deposits in the Toledo District have radically altered Belize's previously untapped mining and manufacturing capabilities.

Guatemala

14,901,286 overall population: Afro-Guatemalan population which makes up about 1-2% of the population *(App. 110,000)*

If one were to look up the population numbers on Guatemala, the most populous country in Central America, many times they will not see a percentage for its Black population. Many Guatemalans, even, are unaware of the fact that there are Black people in the land. Mestizos, Mayan Indians and Black Africans make up the bulk of Guatemalan people. Africans arrived as slaves around the same time that Guatemala was invaded by Pedro de Alvarado around 1524. Sugar plantations and haciendas were the big reason for slavery. Most of the Africans, who were enslaved, eventually intermarried with the Native American population. Slave importation did not last a long time as the conquistador became very nervous due to the uprising of the Blacks they captured. Slavery became less important to the conquistadors and slavery was abolished in 1823. The Afro-Guatemalans one may meet today are the Garifuna peoples who came from Honduras in the 1800s and maintain many of their traditions in art, food and music. The next batch are Afro-Caribbeans (BlackWest Indians) and they speak Creole English as well as Spanish, and mostly have English last names.

Honduras

Black Population: App. 350,000

Nicaragua

Overall population of 6,465,501 (Black Population: App. 570,000)

Mestizos, Blacks, Native Americans make up the bulk of the people but there are also Whites, Arabs and Asians. Besides the Garifuna peoples, Afro-Caribbeans and Miskitos make up the other major African groups in the nation. Slavery began in 1524 and ended in 1821 as Blacks were primarily used for farming purposes, replacing the murdered Natives. Most Blacks live along the Atlantic coast.

Miskitos are located along the Atlantic, or "Mosquito Coast", side of the country and are descendants of Blacks and Indians. The Blacks are descendants of escaped Caribbean slaves. The Mosquito Coast was a region of Nicaragua that was not colonized by Spain, but instead became a British "protectorate". Because of this English is mostly spoken by the people living there. During the civil war, many Miskitos were displaced from their homes by the Sandinista guerillas. Most fled into Honduras but eventually came back in the mid 80s. There are about 75,000 Miskitos in the country. Although, they are of African/Indigenous ancestry, they are mostly associated with their Indian culture because of they have retained their language and culture.

The Garifuna peoples came from Honduras in the 19th century and are located in Orinoco, the Bluefields and Puerto Cabezas. The still speak their African tribal languages and also have a faith-healing festival called Gara-Wala.

Afro-Caribbeans are the largest group of Black people. They arrived as slaves with the British and the Dutch in the 17th century from the British-influenced West Indian island countries, namely Jamaica and the Cayman Islands. They still speak Creole English but the entire population speaks Spanish. The United Fruit Company provided jobs for Afro-Caribbeans during the early 1900s. The Bluefields and Puerto Cabezas have the largest Afro-Caribbean community. "Mayo Ya", an annual festival in May, fuses elements of their West Indian tradition such as Reggae music dancing. Most of these English-speaking Blacks are very educated and hold an edge over the indigenous people and Mestizos in the area.

Costa Rica

Overall population of 4,999,441 (Black Population: Est. 150,000)

Costa Rica's Black population is the largest "minority" in the country. Slavery brought is the first wave of Blacks but more migrated, along with other ethnic groups such as Italians, to become workers on the Costa Rica Railroad and fruit plantations in the late 1800s. The Black population are descendants of Jamaica, Barbados and coastal Africa. The official language of Costa Rica is Spanish but the Blacks speak English as well from their days in the West Indies. Limon is where most Blacks are located. Segregation was a daily ritual for the Blacks who worked on the railroad and banana plantations. The government felt that they were not citizens of the nation so most of the country was off limits to them, sort of like a color bar.

The Afro-Costa Rican population has declined as most moved to neighboring Panama and to the USA but they have attained high education standards are employed in leading professions. Their culture has also been attained as they speak Creole English, practice African religions, perform Caribbean music and enjoy West Indian cuisines. It is because of this that they have not been fully accepted as "Latinos" by Costa Rican Mestizos and Whites because they have not fully adapted to the culture, although they speak spanish. Racism still exists but it is extremely quiet.

Limon is the place where you will find most Black folk. It is still the main port for bananas for Costa Rica. Also, it is home to the best carnival in Central America which was started by Alfred King and takes place in October.

Panama

Overall population of 4,176,869 (Black Population: App. 600,000)

Panama was the first place in the Western region's mainland that had a Black settlement. Formerly, a part of Colombia until its independence in 1903, Panama is not always considered a Central American nation, historically at least. The first Blacks arrived around 1513 as explorers who built vessels, the next batch arrived a few years later as slaves who transported goods from ships and to work on gold mines. The first African slave rebellion in the Americas took place in Panama as they overpowered the slavemasters and received help from the AmerIndians. These people were called "cimmarones" (the wild ones) but are now known as "Playeros" (the beach people), Spanish speaking and Roman Catholic Black people.

1849 marked the building of the Panama Railroad and the opportunity for work. It also marked a second coming of Black people as Afro-Caribbeans, mostly from Jamaica, Barbados and Trinidad, were recruited to work on the railroad. In 1880, the French started work on the Compagnie Universelle du Canal Interoceanique. Its purpose was for a transoceanic canal across the ithmus. Finally, the building of the Panama Canal by the USA began in 1907.

Like the building of the railroad, Blacks were recruited to work for the French and the Americans in Panama. Workers lost their lives during construction of all 3 projects and after the jobs were done, most Blacks remained. Racial segregation has been taking place ever since the building of the canal. A "Gold" and "Silver" label was used in Panama, White workers were paid in gold while Blacks were paid in silver. Public facilities were labeled "gold" and "silver".

The original Blacks in the country are nicknamed "nativos" while the Afro-Caribbeans are known as "antillanos". The lack of unity between these two groups is very surprising. There are still laws that are directed towards Afro-Caribbeans in Panama but they are getting closer everyday to equal human rights.

Both groups have been fighting for their rights for 500 years.

A small example of the thousands of Centra; America's Black people in Science and Technology:

Benito A. Sinclair, P.E., Civil Engineering Co. Exec., Structural Engineering

Benito Sinclair was the first African American to complete the Program at the School of Architecture and Environmental Design at the California Polytechnic State University at San Luis Obispo in 1957, and a few years later, the first graduate from this school to be designated as a Distinguished Alumnus. He thereafter launched into his career as a structural engineer with a couple of private engineering firms before initiating his own professional practice in 1965: Benito Sinclair-Darshville & Associates. In 1968, Mr. Sinclair became the first African American to pass the California Structural Engineers Examination. Initially his firm worked on small projects, but later became a major designer in large projects such as a Los Angeles Subway System. Later, his firm designed 13+ stations along the Los Angeles-Long beach Light Rail System. His firm joint ventured to design the Tom bradley International Terminal at LAX. Mr. Sinclair co-founded the Los Angeles Council of black Professional Engineers (LACBPE), served on the Committee on Minorities at The National Academy of Engineering of the National Academy of Science and has been a Board Member of MESA.

SOUTH AMERICA

Brazil

Black Population: App. 100+ Million

When gold was found in Brazil in the 1690's, this country was finally recognised for its mineral and trading potential. Approximately a century later, though, it was clear that the gold deposits were limited and that the agricultural value of this country remained its main asset. Napoleon Bonaparte arrived in 1807 and the Prince Regent, Dom Joao, arrived shortly thereafter. When Dom Joao returned to Portugal in 1821, he left Brazil in the hands of his son, Dom Pedro. However, when the king tried to return to what was, essentially, his territory (Brazil), his son rebelled, declaring this country's independence from Portugal.

Slavery was a major trend in Brazil, although this was rarely recorded in the official annals of history. These slaves were brought to South America from Africa. Therefore, many of the modern-day people of Brazil have African genes too.

The first permanent Portuguese settlement was established at São Vicente in 1532. Salvador was founded by the Portuguese in 1549 as the first capital of Brazil, and it became a major port for slaves and sugarcane.

Brazil abolished slavery in 1888, the last country in the western world to do so, after millions of African slaves had been imported. Coffee was introduced to Brazil in 1720 and by the mid 1800's, Brazil was responsible for half of the world's coffee production. Coffee and sugar became major products of Brazil, giving the locals work and establishing the country within the world's economy.

As nature adapted to the whims of the market, the soil experienced great cycles of monocultural production. Brazil's north-eastern region, especially Bahia and Pernambuco, was the first area to prosper thanks to the sugar planted there, which guaranteed a steady supply of money until the end of the 18th century. From the late 17th century, news of the discovery of gold and diamonds in the region of Minas Gerais sparked a gold rush. This arid, flat shrubland, so unlike the fertile tropics, was given the designation cerrado, denoting a closed or fenced-off zone. Finally, in the 19th century, it was the turn of coffee, referred to as 'black gold', that shifted the economic axis to the south-eastern region: first to Rio de Janeiro, then to São Paulo. It was assumed that the soil had no limits, but that human arms were needed to extract its resources.

Slaves in Brazil fought back in many ways: they killed their masters and plantation owners, fled into the forests, and mounted revolts. Right from the start, they never ceased to negotiate their conditions, fighting for leisure time, the means to support their families, and the right to practise their customs and worship their deities. They also tried to adapt their cultural practices, putting them in mestiço form.

There were also examples of organised insurrections in which the slaves' resistance led to the creation of quilombos, places where those who had escaped would come together. The term originated on the African continent, specifically in Angola, where it designated a kind of military encampment in which warriors would undergo rites of initiation and embrace military discipline. The biggest community of escaped slaves, however – and possibly the one that survived the longest in Portuguese America – was Palmares. This quilombo's original nucleus comprised around 40 slaves who had escaped from a mill in the state of Pernambuco in 1597.

It's no coincidence that, to this day, the version in Bahia – the state most strongly identified with Afro-Brazilians – called candomblé de caboclo venerates the spirits of indigenous ancestors.

Many problems that characterised the past have persisted to the present day. Poverty continues to ravage a significant chunk of the population, and various indicators place the country among the global champions of social inequality.

Colombia

Overall population of 49,500,000 (Black Population: App. 10,000,000)

Afro-Colombians make up 6.68% of the population, almost 3 million people, according to a projection of the National Administrative Department of Statistics (DANE),[1] most of whom are concentrated on the northwest Caribbean coast and the Pacific coast in such areas as Chocó, whose capital is 95.3% Afro-Colombia is considered to have the fourth largest Black/African-descent population in the western hemisphere, following Haiti, Brazil and the United States.

****

The story of Afro-Colombians is like night and day. Blacks were brought to region which would be later known as Colombia in the 16th century during the Atlantic slave trade. African slaves worked on Gold mines, sugar cane plantations, cattle ranches and large haciendas. Afro-Colombians are very visible along coastal Colombia but during the mid 1970s, many blacks migrated to the larger cities. In 1851, slavery was abolished but it would be over 100 years before Blacks would be truly visible. After the emancipation, the Spanish mestizaje (race mixing) movement was an idea that the elite wanted to put in effect to "wipe out" any trace of Africansm by making the nation "lighter". The movement was not totally successful because many Blacks ran for the jungles where they lived with AmerIndians. Not until 1991, after a very strong popular struggle, did the new Colombian Constitution give Afro-Colombians the right to collective ownership of traditional Pacific coastal lands, and special cultural development protection. Even with this new Constitution, Afro-Colombians still have problems facing them but they are slowly but surely gaining "firsts" in their history of being Colombians. Overall, the daily struggle is still there, as witnessed in the state of Choco as Blacks are the prime victims of the 40 year long civil war. The question is just how longer will it take in order for the rest of the world to find out what's happening to this group.

Ecuador

Overall population of 11,822,000 (Black Population: App. 1,1200,000)

Afro-Ecuadorians...yes, there are black people in Ecuador. Africans were brought over in boats from West Africa to work on the coastal areas and food plantations. The first ship, in 1553, that carried the slaves was stranded on the coast of Esmereldas. Then, the African fought off their white captors. 10% of the population is black. The other groups are mainly whites, mestizos, indigenous peoples and Asians. The blacks and the indigenous peoples are the poorest in the nation. According to some, Ecuador is one of the most racist countries in Latin America. Although slavery was abolished in 1821, it did not officially end until 1881. The profusion of racism with Ecuadorian society constrains blacks both in terms of labor and educational opportunities. Many are concentrated into informal labor markets with little to no job stability or security. If you need a comparison of the treatment of Afro-Ecuadorians, think of the way that the Indians are treated in the USA by the whites. Blacks are located in the major cities but are mostly concentrated in the Esmeraldas (La Capital Negra) and Imbabura. Ecuador does not deny the fact that it has a black population but the country tries their best to limit their exposure except when it comes to boosting national pride for Black athletic achievements or using them and Amerindians as tourists attractions

Marimba is a traditional Afro-Latin art form from the Pacific coast of Ecuador and Colombia, consisting of music, dance and theatrical expressions. A perfoming group usually consists of musicians (marimba, drums, other percussion and vocal) and fights for the rights for Afro-Ecuadorians and for all poor people in her native land dancers (3 or more pairs of male and female). The once declining art of marimba is now reviving as the core of resurgent Afro-Ecuadorian culture.

Venezuela

Overall population of 28,887,118 (Black Population: App. 1,200,000)

60,000 Africans were brought to the land of Venezuela in the 17th and 18th centuries to work on *bleep*o plantations. As in other South American nations, blacks dominate the population along the coast. Venezuela is known as having a Cafe Con Leche culture and proven by the fact that over two-thirds of Venezuelans define themselves as mixed race. However, there are many pure black Africans. Not to mention the large mulatto population. Venezuelans sya that there is little to no racism in the country. However, blacks work in poorly paid agricultural or domestic jobs. Power and wealth remains in the hands of the white (Spanish) elite but remains a highly unequal society. Although, the nation is unequal, Venezuela does have blacks working in high government positions. The African community in Venezuela is very African conscious and they even publish a magazine called Africanias.

Barlovento is the Black mecca for Venezuela. It was known among Europe`s chocolatiers for its high quality c@c@o. For 300 years this was one of Venezuela`s greatest sources of revenue, from plantations worked by large numbers of black slaves. Black pride reigns in Barlovento such as Esmereldas in Ecuador and Choco in Colombia. The Venezuelan African community in Barlovento hosted the Second International reunion of the Latin African Family in 1999 with reps coming from Puerto Rico right down to Argentina. The Africans in Venezuela are now playing a prominent role on the international stage whereas previously, they were unseen.

Argentina

2010 the total population was 40,117,096 (Black Population: App. 2,000,000)

Afro-Argentinians are in danger. How you may ask? Well, the Black population in Argentina are at risk of being erased from existence. Yes, Afro-Argentinians are an endangered species. How can a people be endangered? Blacks are not even included on the official census.

Next to Ecuador, Argentina is very racist as well. Lets start at the beginning. Portugese colonizers brought Africans to Argentina around 1630 from Angola. Portugal could not hang onto the land and its slaves due to its conflict with Spain and to protect its claims in Brazil. Farming and servant duties were the primary jobs of black slaves in this territory. The population in the country in the 1700s was almost 50 percent. The decline would start in the 19th century.

Abolition of slavery occured in 1851. However, there were two ways that a black person would be granted freedom before that. Either by manumission or, more frequently, by coartacisn (self-purchase). Most men participated in manumission while women paid for freedom.

The reasons which contribute to the endangerment to Afro-Argentinians are as follows. First, black men were heavily involved in Argentina's wars with Great Britain in 1806-1807, the wars for independence from 1810 to 1816 against Spain, the civil wars throughout the 1820s, and wars against Brazil and the Indian population. Looking to gain social and economic mobility promised by politicians Blacks fought in the Indian extermination campaigns of the 1830s and 1840s. Second, racial intermarriage was encouraged due to the deaths of black men in the wars and for possible social mobility for mixed kids. Third, Argentina's desire to be a European nation in the western region. Aregntina has long been obsessed with the idea of modeling the country after Italy or Spain while making the land whiter while "wiping out" the Natives and the blacks. In Spain, Italy and France, however, there are sizable black communities, but Argentina is obscssed with being a totally White republic. Policies to attract European immigrants worked as a migration began from Europen nations Between 1869 and 1914. a large number of Afro-Argentine women married European immigrants, thereby losing their ethnic identity.

Today, whites make up about 85 percent of the nation and mestizos make up about 15 percent. Blacks are more exoticized than stigmatized but are still kept below the poverty line. Due to the decline, Argentina can deny their African history and the fact they are one of Latin America's most racist nations. A museum worker in Buenos Aires in an interview said "We can't waste space putting things that don't have any relevance to our history". That is a very bold statement when the Tango is a dance and music with such strong West African roots and adored in the country.

Peru

Overall population of 32,919,654 (Black Population: App. 1.5 Million)

As the historian Frederick Bowser describes in his classic study *The African Slave in Colonial Peru, 1524-1650*, African captives in Peru cleared land, laid the streets, carried supplies and built the churches, homes and palaces of the Spanish elite; indoors, they served as cooks, cleaners, nannies and domestic servants. Meanwhile, urban black labor ran much of Lima's daily business; Africans and their descendants worked as artisans, street vendors, bakers, water carriers, gardeners and fruit and vegetable sellers.

In 2009, Peru became the first Latin American country to issue an official public apology to its *afrodescendiente* population for centuries of "abuse, exclusion, and discrimination."

Throughout June, the Peruvian government has been promoting the Month of Afro-Peruvian Culture, with cultural activities and conferences to "make visible", domestically and internationally, the historic contributions of Afro-Peruvians.

As stated by **Omar H. Ali,** there are an estimated three million Afro-Peruvians. This amounts to less than ten percent of the nation's total population—a significantly lower percentage than in the early colonial period. The ending of the slave trade (and therefore new Africans), the migration of indigenous peoples from the highlands to the coastal cities, and pressures to assimilate into the dominant society are all factors for the drop in the visible black population. Adding to this was the increase of new immigrant groups, including Chinese indentured laborers after the abolition of slavery in 1854, followed by Italian, German, Polish, Czech and Japanese immigrants.

A sculpture of San Martín de Porres, who is especially venerated in Peru for his healing powers. Image courtesy of Omar H. Ali. St. Martin de Porres was born in Lima, Peru on December 9, 1579. Martin was the illegitimate son to a Spanish gentlemen and a freed slave from Panama, of African descent.

Uruguay

Overall population of 3,500,000 (Black Population: App. 350,000)

Black Uruguayans are estimated to be about 190,000 and constitutes 10% of the population according to UN and World Bank Reports. They are mainly concentrated in the city of Montevideo. The blacks came to Uruguay as slaves, ladinos--hispanized slaves, in 1534, but eventually settled in Argentina.

A small example of the thousands of South America's Black people in Science and Technology:

Henrique Cunha Júnior, Ph.D.
Electrical Engineering

Dr. Cunha Jr. is currently Coordinator of the Undergraduate Program in Electrical Engineering-UFC. Fellow of African Scientific Institute - (Member) (Source: Lattes Curriculum). Dr. Cunha Jr. experiences in Electrical Engineering include electrical machines, control systems, industrial automation, electrical traction and electrical engineering education. He studied Sociology at the Universidade Estadual Paulista Julio de Mesquita Filho (1979) and Electrical Engineering in the School of Engineering of São Carlos - University of São Paulo (1975). He earned his Masters degree in Electrical Engineering (1980) and MA in History - Université Nancy I (1981) and PhD in Electrical Engineering from the Polytechnic Institute of Lorraine (1983). Dr. Cunha Jr has directed amateur theater groups in the 1970s and was a member of Congada Group, San Carlos. He also participated in the founding of the Association of Black Researchers and was its first President. Dr. Cunha Jr. Also works in graduate and research in Brazilian Education, FACED-UFC, in the following Africanities themes, Afrodescendência, Urban Space, Ethnic Relations and History and African and Afro-descendant Culture.

Rafael Sanzio Araújo dos Anjos, Ph.D.
Transportation Engineering

He holds a degree in Geography from the Federal University of Bahia (1982), a Specialization from the State University of São Paulo (Rio Claro 1985), a Master's Degree in Urban Planning from the UnB FAU (1990), a Doctorate in Space Information at the Dept. of Transportation Engineering by USP (1995) and Post-Doctoral Degree in Ethnic Cartography at the Royal Central African Museum in Tervuren - Belgium (2007-2008). He is currently Professor at the University of Brasília and Director of the Center for Applied Cartography and Geographic Information (CIGA), where he is Coordinator of the Afro-Brazilian Geography Projects: Territory Education and Planning (GEOAFRO Project) and Geographic Instrumentation, Spatial Education and Territorial Dynamics. He was Coordinator of the Postgraduate Program in Geography (Master and Doctorate) and the Milton Santos Geographical Documentation Center (CDGMS) of UnB in the 2010/2011 Biennium and Full Representative of the Institute of Human Sciences at the Council of the University of Brasília in the Biennium 2013-2014. His other lines of research and research are associated with the production of didactic cartographic materials for different levels of education. He coordinates the GEOBAOBÁS / CNPQ Consolidated Research Group, is a member of the Brazilian Committee of the Slave Route.

Cheila Goncalves Mothe, Ph.D.
Chemical Engineering

Every day, thousands of tons of various wastes are dumped in nature, which adds to pollution and consequently the incidence of diseases and health problems in the population. In search of a solution to the problem, chemical engineer, professor and researcher at the Federal University of Rio de Janeiro (UFRJ) Cheila Mothé Gonçalves has been coordinating a study designed to evaluate the process of burning waste from sugar cane, Coconut

Rosa Katemari, Ph.D.
Physics

Dr. Katemari is a university Professor in Physics. When she decided to do Physics, she didn't think about these questions of what would be the representation of women and black people in Physics. She wanted to do Physics since she was 8 years old. The reason is because she loved Astronomy; she loved what I thought Astronomy was. She found it all fascinating. Then, she discovered that many astronomers have a background in Physics. Then she said: This is what she was going to do: Physics. **Dr. Katemari started the research "Telling our story: blacks and blacks in science, technology and engineering in Brazil".**

Farith A Diaz Arriaga
Civil and Environmental Engineering

Farith has been a Red Cross in Columbia, teaching single mothers how to prevent diseases in newborns and infants through the use of potable water in the city of Quibdó. He visited poor communities living around rivers to control the bodyweight of infants, and supplied special food provided by international donors. He taught Physical Chemistry, Design of Drinking Water Plants, Math and Physical-Chemical Processes at Universidad Tecnológica del Chocó, Colombia and coordinated the graduate program in Basic Sanitization. Farith researched on water quality and treatment and Supervised five technicians in the wastewater/drinking water treatment area. He also managed wastewater treatment plants to ensure compliance with governmental regulations. Farith earned his BS degree in Chemical Engineering from Universidad de Antioquia (Colombia), and his MS degree in Civil Engineering from Universidad de los Andes (Colombia). Currently, Farith is now in USA to earn his Ph.D. in Environmental and Water Resources Engineering,

 **Roland King**
Civil Engineering

Roland King has more than 20 years Management Professional Experience. His expertise includes: Project management of drainage of polders on tidal rivers in relation to the optimal production of crops, road building and Master plans of Roads and Waterways (Infrastructure) of Areas for Water control. Design of Sluicegates, Water pumps and pumping stations, Channels, conduits etc.; Drafting Terms of References and tender documents; Financial Economic Analysis of projects: Cost estimations, Calculation of Internal Rate of Return (IRR), Benefit-Cost (B/C) ratio, Net Present Worth value en de Net Benefit-investment (N/K) ratio). Cash Flow Statements, Income statements, Balance Sheets, Among others Accountancy Ratio's: Quick Ratio's, Solvability etc.; Development of business plans; Project engineering, Financial and technical analyses of Agricultural Projects or companies with the aim of enhancement of their efficiency & effectiveness for better profits; *Water management" and "Project analyses/evaluation Land design" of the University of Suriname*; Design of Sluice gates, Sluices, conduits, Weirs, Pumping stations, canals, master plan of roads & canals, Masterplan of the Projects: Emplacements, Roads & Canals, Communities, working area; Design and Construction of drainage and irrigation dams. He has worked in Suriname, The Netherlands, Mozambique.

****

CARIBBEAN

Cuba

Overall population of 11,327,291 (Black Population: App. 2,500,000)

Unlike other Latin American countries, or in North America for that matter, racial discrimination in Cuba has mostly been non-violent, mostly verbal. One known incident of violent racism was in 1912 when government troops killed about 3,000 blacks in fighting that erupted after an Afro-Cuban political party was declared illegal. One Afro-Cuban is quoted as saying "There is no official racism here anymore but there is still a culture of racism. The mistake was to think that just by having everyone integrated, racism would fade away." One of the myths is that there is no discrimination in Cuba. Sure Blacks, Whites, mixed people, Asians and others interact with each other but racism still persists. Work is being done to bring about changes, though. Christopher Colombus landed on the island in 1492. After the decimation of the AmerIndians, Africans were brought in as slaves to work on sugar plantations. The fact that sugar was the basis of the Cuban economy, many more Blacks were enslaved to work on the island's crops. Slavery was abolished in 1886, one of the last nations to do so. Upward mobility for Blacks has improved considerably, although they are still underrepresented in the high levels of government and the communist party. Blacks have particularly found advancement in military careers and in Cuba's highly successful sports programs. If there ever were a spanish-speaking nation that epitomized African pride it would be Cuba. As a matter of fact, the Marxist government of Cuba has declared Cubans an Afro-Latin American people and has formed close ties with Angola, Ethiopia, and other African states. Most Afro-Cubans are proud of their blackness and consider Cuba to be a "Black country".

Jamaica

Overall population of 2,959,569 (Black Population: App. 2,500,000)

Columbus landed on 14 May 1494, on his second American voyage of exploration. He named the island Santiago (Saint-James). However, the name was never adopted and it kept its Arawak name Xaymaca, of which 'Jamaica' is a corruption. Lacking gold, Jamaica was used mainly as a staging post in the scramble for the wealth of the Americas.

The Spanish arrival was a disaster to the indigenous peoples, great numbers of whom were sent to Spain as slaves, others used as slaves on site, and many killed by the invaders, despite the efforts of Spanish Christian missionaries to prevent these outrages. There were no Arawaks left on the island by 1665, but there were enslaved Africans replacing them.

In 1645 the British captured Jamaica from the Spaniards, whose former slaves refused to surrender, took to the mountains and repelled all attempts to subjugate them. These people came to be known as Maroons

Settlers, using slave labour, developed sugar, cocoa, indigo and later coffee estates. The island was very prosperous by the time of the Napoleonic wars (1792–1814), exporting sugar and coffee; but after the wars sugar prices dropped, and the slave trade was abolished in 1807.

The island is the birthplace of Rastafarianism and the movement played a tremendous role in 20th century Jamaica. The Jamaica born leader **Marcus Garvey**, who led the *United Negro Improvement Association*, encouraged people to "Look to Africa," where he predicted a black king would be crowned, who would serve as a redeemer. Soon thereafter, Haile Selassie was crowned emperor of Ethiopia; the word Rastafari comes from Selassie's name at birth, Tafari Makonnen, and the word "Ras," meaning "prince."

Francis William: 1702 – 1770, Jamaican Mathematician

Mathematics Degree -Cambridge University

Francis Williams was selected to take part in a social experiment devised by the Duke of Montagu who wished to show that black individuals – with the right education – could match the intellectual achievements of whites.

Challenge
A rival once suggested-"The abstruse problems of mathematical institution turned his brain; and he still remains, I believe, an unfortunate example, to show that every African head is not adapted by nature to such profound contemplations."

Achievements
Degree in Mathematics, Latin and Literature from Cambridge University. He returned to Jamaica to set up a school, teaching Mathematics, Latin, Reading and Writing.

Haiti

Overall population of 11,385,338 (Black Population: App. 11,000,000)

When Columbus landed in the island of Hispaniola on December 6, 1492, he found a kingdom ruled by a cacique, or Taino Indian chief. After the French arrived in the seventeenth century to continue European exploration and exploitation in the Western Hemisphere, the indigenous population was largely exterminated. As a result, Africans (primarily from West Africa) were imported as slave labor to produce raw goods for international commerce. Considered France's richest colony in the eighteenth century, Haiti was known as "the pearl of the Antilles." Resisting their exploitation, Haitians revolted against the French from 1791-1804. One of the most important outcomes of this revolution was that it forced Napoleon Bonaparte to sell Louisiana to the U.S. in 1803, resulting in a major territorial expansion of the United States. This land deal doubled the size of the U.S., adding to its holdings either in part or whole: Louisiana, Arkansas, Nebraska, Missouri, Iowa, Oklahoma, Kansas, Minnesota, the Dakotas, Colorado, Wyoming, and Montana.

When Haitians took their independence in 1804, they changed their colonial name from Saint Domingue (the name given by the French) to its Taino name of Haiti, or Ayiti in Kreyòl.

The United States occupied Haiti from 1915-1934, changed Haiti's constitution, and in many ways further contributed to its ongoing instability, many African-Americans denounced the occupation of a sovereign nation.

From San José State University, USA, Department of Economics

Compared to the political history, the economic history of Haiti is relatively simple. The original economic basis for the Spanish colonies on Hispaniola was sugar plantations. The French continued the sugar economy and introduced coffee. There were other plantation crops grown such as cotton and cacao for chocolate but it was sugar and coffee that were the most important. Under the French plantation system, based upon slave labor, Haiti was an enormously profitable operation. The Haitian sugar economy was in competition with the northeast region of Brazil, which previously had been the major source of sugar for Europe. The French sugar and coffee operations in Haiti were so productive that its exports to Europe were comparable and perhaps exceeded the total exports of the British North American colonies.

After the battles associated with independence there was some attempts to retain the large scale plantation agriculture of the colonial period but that effort was doomed. Land was distributed into small scale farms but these units devoted only a fraction of their resources to growing export crops like sugar and coffee. Often the output is consumed domestically and there are no exports of sugar or coffee.

In the latter part of the 20th century tourism became an important element of the economic base of Haiti. But the political instability and the public's association of Haiti with AIDS severely crippled the Haitian tourism industry.

In recent decades the low wage rates of Haiti have attracted manufacturing assembly operations. Haiti is one of the few countries that has pay scales low enough to compete with China.

Dominican Republic

Overall population of 10,630,000 (Black Population: App. 8,000,000)

Afro-Dominicans or Dominicans of African ancestry, are Dominicans whose ancestry ties within the continent of Africa. Most of them came from West Africa and the Congo. The first Africans in the Dominican came in 1502 from Spain, 8 years later African-born slaves came in large numbers.

Since 1492, when Columbus landed on Hispaniola, the Dominican Republic has seen wave after wave of foreign interlopers. The destruction of the native population led to periods of neglect and conflicts between the French and Spanish colonial systems.

As the Taíno civilization collapsed, so did the gold mines, and no amount of imported African slaves could make up the shortfall. Spain dropped Hispaniola as quickly as it had found it, turning its attention instead to the immense riches coming from its new possessions in Mexico and Peru

The Dominican Republic's road to independence has been shaky. The territory briefly reverted to Spanish rule during the 1860's and was twice occupied by the United States. The first lasted from 1916 to 1924, while the second was from 1965 to 1966 and several volatile dictatorships ruled in between these periods.

From Wikipedia

The Dominican Republic's most important trading partner is the United States (about 40% of total commercial exchange). The country exports free-trade-zone manufactured products (garments, medical devices, gold, nickel, protection equipment, bananas, liquor, cocoa beans, silver, and sauces and seasonings. It imports petroleum, industrial raw materials, capital goods, and foodstuffs. In 2005, the Congress of the Dominican Republic ratified a free trade agreement with the U.S. and five Central American countries, the Dominican Republic – Central America Free Trade Agreement (CAFTA-DR). The total stock of U.S. foreign direct investment (FDI) in Dominican Republic as of 2006 was U.S. $3.3 billion, much of it directed to the energy and tourism sectors, to free trade zones, and to the telecommunications sector. Remittances were close to $2.7 billion in 2006

Puerto Rico

Overall population of 3.194 million (Black Population: App. 2.427 million)

March 22, 1873 marked an important day in Puerto Rico's history. It was the day that slavery was finally abolished. Puerto Rico, nicknamed the "Island of Enchantment" was, for Blacks and Indians an "Island of Disenchantment" or an everyday living hell during the colonial period. As in other western region nations, the Amerinidians mostly were either worked to death or died from diseases caught from the conquistadors. Puerto Rico, which translates into "rich port", became the new home for many Africans as they were forcefully brought over to take the place of the declining Tainos to produce coffee, tobacco and most importantly sugar. From the day they set step through the port around the year 1560, Afro-Puerto Ricans were seen as inhuman, strange, exotic and suspicious. Slavery was an expensive business to be engaged in so White slavemasters began branding the foreheads of slaves to distinguish legal slaves from the illegal ones and keep them from being kidnapped by rivals. Just like the U.S.A. during slavery time, Puerto Rico had what is known as "house negros" and "field negros".

If it wasn't for Ramon Betances and the rest of the Puerto Rican Abolistionist movement, slavery came to an end after over 350 years. It was definitely a business decision on behalf of the Spanish National Assembly as they were compensated with 35 million pesetas per slave and Blacks had to work three (3) more years before totally losing the shackles. The abolition of slavery on the island did have another cost. Just as in Mexico, Puerto Rico according to many scholars suffers from "African Amnesia". Take into fact that in 1820, Blacks made up 56% of the nation and then in 1950 it was down to 23%. In the 2002 census,

8% out of 3 million plus people identified as Black while 80% said they were White. It is without a doubt that Puerto Rico is "the lightest country in the West Indies" but one can't only determine race by only skin color as is done on the island. Truthfully speaking, Puerto Ricans are more Mulatto than Mestizo plus there also is a small surviving Amerindian Taino population even today. Blackness is acknowledged, though, but rarely in a positive light. The ones who do acknowledged it are mostly Black or dark-Mulatto and have been seen as less "Puerto Rican" due to the racial policies still practiced in PR and the relationship with the U.S.A. has played a huge part in the denial of Blackness. Even though native Boriquas like Ricky Martin and Mark Anthony have embraced the African element in Salsa music, Black history in Puerto Rico exists but it does not exist. However, thanks to some, Puerto Rico is starting to regain its memory letting the world know that there was an African past therefore influencing the present. Maybe one day, Puerto Rico can truly be the "Island of Enchantment". *from the upcoming book, Afros All Over*

Trinidad-Tobago

Overall population of 1,328,018 (Black Population: App. 520,000)

The Spaniards subdued and enslaved the native Caribs and Arawaks but until the late 1700s paid little attention to Trinidad as other ventures were more profitable. As a result, Trinidad's population was only 2,763 in 1783. Amerindians composed 74 percent of that total (2,032). Although African slaves were first imported in 1517, they constituted only 11 percent of the population (310) in 1783. Indeed, the slave total was barely larger than the 295 free nonwhites who had emigrated from other islands. The remaining 126 Trinidadians were white.

In an effort to make Trinidad more profitable, the Spanish opened the island to immigration in 1776 and allowed Roman Catholic planters from other Eastern Caribbean islands to establish sugar plantations. Trinidad linked landownership to the ownership of slaves; the more slaves, the more land. Land grants were also given to free nonwhite immigrants, and all landed immigrants were offered citizenship rights after five years. As a result of this new policy, thousands of French planters and their slaves emigrated to the island in the 1780s and 1790s. By 1797 the demographic structure of the island had changed completely. The population had expanded dramatically to 17,718, about 56 percent of whom were slaves.

The British, who were at war with Spain and France, conquered Trinidad in 1797 during the Caribbean unrest that followed the French Revolution. Trinidad was formally ceded to Britain in 1802. A decade after slavery was abolished in 1834, the British government gave permission for the colonies to import indentured labor from India to work on the plantations.

By the late nineteenth century, Trinidad and Tobago were no longer profitable colonies because sugar was being produced more cheaply elsewhere. In 1889 the British government united Trinidad and Tobago in an effort to economize on government expenses and to solve the economic problems of the islands.

Sugar remained the backbone of the colonial economy throughout the nineteenth century, but cocoa expanded rapidly towards its end, and overtook sugar as the colony's leading export crop by about 1900. Both crops dominated plantation production into the 1900s, but cocoa declined sharply in the 1920s to 1940s, never to recover. The exploitation of oil, found in the southern half of Trinidad, became the colony's most important source of revenue by the 1930s. The demands of World War 2, and the development of marine production in the 1950s, only increased the domination of oil in the island's modern economy.

Today, the economy of Trinidad and Tobago is the wealthiest in the Caribbean and the third-richest by GDP (PPP) per capita in the Americas. Trinidad and Tobago is recognised as a high-income economy by the World Bank. Unlike most of the English-speaking Caribbean, the country's economy is primarily industrial,[14] with an emphasis on petroleum and petrochemicals. The country's wealth is attributed to its large reserves and exploitation of oil and natural gas

Physician, sociologist and politician **José Celso Barbosa** was one of the first Puerto Ricans and persons of African descent to receive a medical degree in the US. But he also made history in plenty of other ways. He served in the executive cabinet under Puerto Rican Governor Charles H. Allen and joined the first Puerto Rican Senate. Barbosa, who advocated for statehood, also established *El Tiempo*, the island's first bilingual newspaper. (Written by Jenay Wright)

A small example of the thousands of Caribbean's Black people in Science and Technology:

 **Bert Fraser-Reid, Ph.D.**
Chemistry

Distinguished Research Chemist and James B. Duke Professor (rec'd by only 43 out of 1400 professors) Emeritus, has written over 350 publications. He has lectured at universities, academies, companies, and prestigious institutions in 47 countries. He synthesized insect pheromones from glucose, and shown that many synthetic petroleum based products can be made from sugars. His lab has discovered reactions to make complex sugars, known as oligosaccharides, which are among nature's most important biological regulators, particularly for the body's immune system. He retired from Duke in 1996 and founded a private non-profit research Institute, with a goal to develop carbohydrate-based therapeutic agents for Third World infectious diseases, under the sponsorship of the World Health Organization. In May, 2000, Dr. Fraser-Reid was chosen as the only US member of a Consortium of six international interdisciplinary scientists, funded by the prestigious "Human Sciences Frontier Programme Organization" of Europe, to work towards a carbohydrate-based anti-malaria vaccine. His Institute accomplished the first syntheses of antigenic oligosaccharides associated with malaria and tuberculosis. Dr. Fraser-Reid has won the world's *premiere* award in Carbohydrate Chemistry, also national chemistry awards from societies in USA, Canada, Japan, the Alexander von Humboldt Senior scientist Award from Germany.

Ray M. S. Brathwaite
former Chairman, Solid Waste Management Co.; Oil Industry Development

For more than 15 years, Mr. Brathwaite has been Principal interventionist providing services to facilitate business turn-around. Offering expertise in Performance & Process Improvement; Situational Leadership Development; Business Development Strategies; Change Management; Performance & Consequence Management Systems; MBTI Facilitation; et al].

Cardinal Warde, Ph.D.
Electrical Engineering and Computer Science; President of the Caribbean Diaspora for Science, Technology and Innovation, Interim Executive Director of the Caribbean Science Foundation

Dr. Warde specializes in the field opto-electronics - the branch of electronics that deals with electronic devices for emitting, modulating, transmitting, and sensing light. He is one of the world's leading experts on materials, devices and systems for optical information processing. He is currently working on optical computing applications such as opto-electronic neural network processors, which have brain-like computer functions and hologram technologies. This innovator is the holder of over 12 patents. Dr. Warde's most exciting invention is a unique pair of "display glasses" which are spectacles with a tiny liquid display unit attached. It is said to be the smallest sharpest display unit ever devised, and it allows a holographic image to be displayed 18 inches away from the eyes. Dr. Warde is currently working on building computers with brain-like functions and face recognition so they can make "judgement calls" when presented with incomplete information. He is also collaborating with NASA to develop software with space imagery applications. He is the co-inventor of the Microchannel Spatial Light Modulator, Membrane-Mirror light shutters based on Micro-Electro Mechanical Systems (MEMS).

 **Claire A. Nelson, Ph.D.**
Futurist; Sustainability Engineer

Dr. Nelson is Chief Ideation Leader, The Futures Forum. She has been actively engaged in the business of international development for over thirty years, working in the area of project development and management. The first Jamaican woman to hold a Doctorate degree in an engineering discipline and only Black in her graduating class. Dr. Nelson holds Industrial Engineering Degrees from the State University of New York at Buffalo, Purdue University, and a Doctorate in Engineering Management from the George Washington University. She is also recognized as a Social Entrepreneur and is the Founder and President of Caribbean American advocacy organization, Institute of Caribbean Studies, and Architect of National Caribbean American Heritage Month. She has served on numerous Boards and Committees including: Sustainability Division, Institute of Industrial and Systems Engineering; World Futures Review Journal; TechCast; AAAS Fellowship Review; US Department of Commerce US/Caribbean Business Development Council; Commission on Pan-African Affairs, Office of the Prime Minister of Barbados; African-American Unity Caucus. She is recognized as one of top 100 Women Futurists, and was honored as an Outstanding Alumni of Purdue University; and a White House Champion of Change.

 **Gerald Lalor, Ph.D.**
Director-General of the International Centre for Environmental and Nuclear Sciences (ICENS); Chemistry

Hon. Dr. Lalor began his career at the University of the West Indies in 1960. He became a Professor of Chemistry in 1969 and head of the Department of Chemistry between 1969 to 1972. Professor Lalor was promoted to Pro-Vice Chancellor of the University in 1974. He was appointed Principal of the University in 1991 and served in this position until 1996. Upon retiring from his position as Principal, he spearheaded the formation of the International Centre for Environmental and Nuclear Sciences (ICENS). As Director of ICENS, Dr. Lalor acquired the use of the Slowpoke Nuclear Reactor for research; and led a research team that developed a geochemical map of elements in Jamaican soils which aids in the detection of contaminated land and provides important information for government planning agencies. Dr. Lalor has been awarded the Order of Jamaica, the Sir Philip Sherlock Award for Excellence, Commander of the Order of Distinction, and the Gleaner Annual Award for his work in the development of the University of the West Indies Distance Teaching Experiment (UWIDITE), and the Norman Manley Award for Excellence and the Musgrave Gold Medal for his contribution to the development of Science and Technology in Jamaica.

Jose Carlos Lorenzo Feijoo, Ph.D.
Agricultural Sciences

Dr. Feiloo has been permanently employed at the Bioplant Centre (University of Ciego de Avila, Cuba) from September 1992 to the present. He has over 20 years of experience in teaching, designing, implementing and managing fundamental and applied research activities in agricultural sciences, with strong background in biotechnology. Dr. Feiloo has more than 40 publications documented (763 citations) and Member of the Cuban Academy of Sciences. He has always been Interested in continuing research in the use of biodiversity in plant breeding and use of biotechnology as a tool. In 2001, he became the head of the Laboratory for Plant Breeding. The main research topics he has been involved in are: • Plant micropropagation in temporary immersion bioreactors (sugarcane, pineapple, ornamental plants). Physiological roles of phenolics (sugarcane). Production of proteases (pineapple). • Plant genetic transformation (citrus, pineapple). • Protoplast culture, somatic embryogenesis and artificial seed technology (citrus). • Development of early selection systems of plant resistance to biotic stresses (pineapple-Phytophthora, banana-Fusarium). • Cryopreservation of plant germplasm (sugarcane). • Use of biostatistics in plant biotechnology.

EUROPE

The 2011 census found that 1.85 million of a total **Black population** of 1.9 million lived in **England**, with 1.09 million of those in London, where they made up 13.3 per cent of the **population**, compared to 3.5 per cent of **England's population** and 3 per cent of the **UK's population**.

As of 2013, there were **approximately 817,000 Afro-Germans in Germany** with a population of 82,000,000 people. This number is difficult to estimate because the German census does not use race as a category, following the genocide committed during World War II under the "German racial ideology". German-born blacks are sometimes called *Afrodeutsche* (Afro-Germans) but the term is still not widely used by the general public. This category includes people of African heritage born in Germany. In some cases, only one parent is black.

In May 2013, France's National Assembly successfully voted on a bill to remove the words 'race' and 'racial' from the country's penal code. But critics were quick to point out the disparity between constitutional reform and actual practice. Between one and five million French citizens claim African or Caribbean heritage. These numbers are, however, estimates, as population censuses do not recognize race. Statistical data indicates **approximately 2 million Black people live in France** . Generally speaking, the Black people represent approximately 3.5% of the population in France.

A small example of Europe's Black people in Science and Technology:

 Halima Moncrieffe, Ph.D.
Immunology

Dr. Moncrieffe's research aims to understand why people get autoimmune diseases like arthritis and lupus, and to find the best therapy for each individual using precision medicine. A research highlight includes identifying a new potential protein trigger of juvenile idiopathic arthritis, a disease that affects about 1 in 1000 children. This CCHMC-Albert Einstein College of Medicine collaboration resulted in a JCInsight publication and numerous media reports. As project manager for several national and multi-national studies, she contributed to moving the field forward with domain expertise and peer-reviewed publications in the fields of rheumatology, precision medicine, immunology, genetics and healthcare disparities. Sharing knowledge to "build as I climb" including management strategies and technical training. she educates the next generation of medical practitioners and scientists and take pride in seeing their awards and increased critical thinking, communication and analytic skills.

 **Donald Palmer**

Donald Palmer, born in 1962 in London, is Senior Lecturer in Immunology in the Comparative Biomedical Sciences Department of the Royal Veterinary College, London. His research is concerned with the role of the thymus in the ageing of the immune system, and the identification of 'novel markers' on the surfaces of cells. Donald's parents emigrated from Jamaica in the early...

 **Mark Richards**

Mark Richards, born in 1970 in Nottingham, is a Senior Teaching Fellow in the Department of Physics, Imperial College, London. His research into wireless air sensing networks for real-time pollution mapping is linked to his role as co-founder and director of Duvas Technologies Ltd. Mark's parents emigrated from Jamaica in the 1960s. Following secondary and further education at a comprehensiveschool in Nottingham, Mark read chemistry at the University of Manchester, 1988-1991. He worked as Technical Sales Officer in spectroscopy for Perkin Elmer and then completed a PhD at Imperial College on 'High Resolution Spectroscopic Analysis of Trace Atmospheric Gases', 1994-1998. His next post in science in 2002 was as a post doctoral researcher in an Imperial College 'Technology Transfer Program', leading to the formation of Duvas Technologies in 2008. Mark is involved in a number of equal opportunities and schools 'outreach' activities and was involved in Tony Sewell's 'Generating Genius' programme. He also DJs and remixes as DJ Kemist, with his own independent label Xtremix Records.

ASIA

Southeast Asia

A small example of Asia's Black people in Science and Technology:

Sharala Axryd - Founder & CEO, Center of Applied Data Science

With a passion for data science and over 15 years of experience in the telecommunications field under her belt, Sharala Axryd is leading the data-driven business transformation and driving the benchmark for data science education in the ASEAN region. A thought leader in the data science space, she is a highly-sought after speaker for conferences with topics ranging from analytics to women in STEM. Award-winner of the EY Woman Entrepreneur Of The Year 2017 Malaysia, SEBA 2018 Woman Technopreneur of the Year and among the Digerati 50 by Digital News Asia (DNA), she is the Founder and Chief Executive Officer of The Center of Applied Data Science (CADS), ASEAN's first and only one stop platform and center of excellence for Data Science. She was part of the team that brought in The Data Incubator (an American-based data science center) to Malaysia, launched ASEAN's first data science accelerator program in 2016 and spearheaded an initiative with the Harvard Business School in Boston to support Malaysia's national agenda to be the hub for Big Data Analytics (BDA). Prior to CADS, she worked with a diverse set of clients to deliver hands on training through several workshops and trainings on GPRS Performance Analysis and Optimisation. As the Founder and Managing Director of ULearn, Sharala and her team was successful with the Hands On Technology Training (HOTT), eventually evolving to be the first in the industry to develop an Automated Competency Gap Analysis (UrSkillsReporter).

Estelle Cooke-Sampson, Korean War baby

Dr. Estelle Cooke-Sampson, MD is a Diagnostic Radiology Specialist in Washington, DC and has over 42 years of experience in the medical field. She graduated from Georgetown U, School of Medicine medical school in 1978. Be sure to call ahead with Dr. Cooke-Sampson to book an appointment.

Future

Sustainable Industrialization will drive economic development. Products and services must be able to service domestic consumption, while simultaneous providing these products and services for global consumption. Just recently, an interview of Dr. Peter Onwualu (an ASI Fellow) said 'Only manufacturing can change Nigeria'. He suggest establishment of a Foundation for Industrial Development that provides a mechanism that will enable research institutes to be able to have the kind of funding they require to be able to do the kind of research that can impact on industries.

I attended a joint session called by President Clinton in the 1990s that brought together government agencies, academic, and industrial resources to address how brainpower could more efficiently be used and shared to help the U.S. move forward. Then Commerce Secretary Ron Brown headed the U.S. delegation. I presented Secretary Brown with one ASI's *"Blacks In Science Calendar"*, which was then presented to then President Clinton's Science and Technology Advisor. Two people listed in our calendar then became science and technology committee advisors to the Whitehouse. I am suggesting that all of our resources collectively be focused to help us move forward.

Africa must learn to develop with its resources and not rely so heavily on loan and donations from foreign donors like the IMF and World Bank. Africa has resources that the world needs. The game is to keep Africa de-stabilized so its resources can be plundered at little to no cost.

How can countries that are so small as those in Europe control the economic development of Africa? Divide and conquer. *The "Willie Lynch speech (Willie Lynch in 1712, presented a speech of his doctrine: "there are many ways in which you can keep control of your slaves".* Saying in another way: "Keep the body strong and the mind weak") practices are still being performed today. Money is given to Black athletes and entertainers, but not to scientists, engineers and educators. Images of success are presented in the media as being athletes and entertainers. So our children want to follow in the footsteps of athletes and entertainers. In Africa, we know all too well what is happening to keep our countries unstable. "You cannot have a successful thief unless you have an equal culprit who buys that which is stolen". Money is stolen, but where is it located? If you must steal, build something! Take the U.S. robber barons around the turn of the 20th century[1,2].

Education is key to Africa's development; but how do you retain brainpower when the brain drain is so prevalent in Africa? Vocational skills are needed as much as skills acquired from universities and colleges. Booker T. Washington and Dr. John Hope/W. E. B. DuBois argued about what was most important: but I say they are simultaneously required.

We of the African Diaspora must join hands. We must seek to move our youth forward by developing "individual development plans" for each child to help them aspire towards their dreams. "Each One Teach One". Pass on positive news and images (commercial media usually does not). Learn how to become proficient in using the internet. Keep up with emerging ideas and technology.

****

We'll only stop young African scientists from leaving if we help them fix these key problems at home

October 2, 2018

By Anna Coussens, Abidemi James Akindele, Badre Abdeslam, Fridah Kanana & Mona Khoury-Kassabri

Young African scientists face persistent barriers which cause them to leave their own countries, and even academia. This means the continent's work force loses highly trained people who are crucial for scientific and technological advancement, and for economic development.

It's estimated that 20,000 highly educated professionals leave the continent annually, with up to 30% of Africa's scientists among them.

A number of factors contribute to this trend. The extreme factors include war and political instability. But the more common "pushes" are a desire for higher pay, better opportunities, and the search for a conducive research environment – one where infrastructure and management help drive careers and research potential.

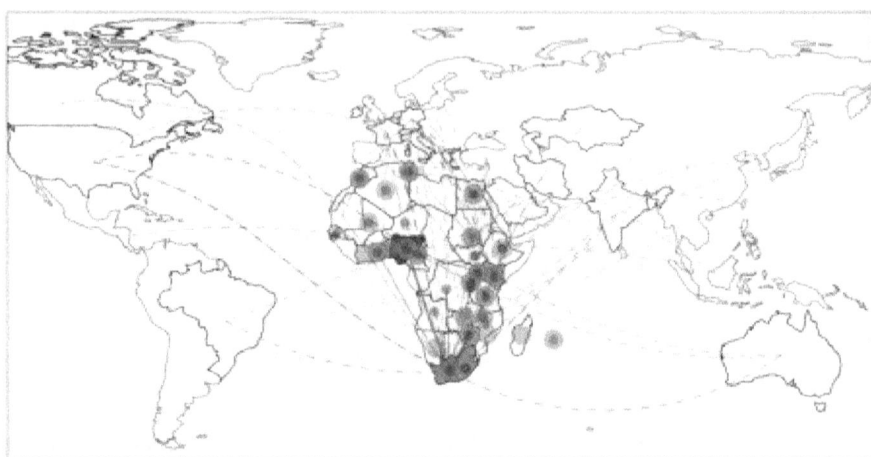

Most young African scientists leave for higher pay, better opportunities, and a conducive research environment.

To identify all the barriers and develop strategies to address them, the Global Young Academy – an organisation of 200 talented young scientists and over 200 alumni from 83 countries – established the Global State of Young Scientists Africa project. Working with local research partners and international higher education experts, the project aims to identify the challenges and motivations that shape young scientists' career trajectories. Our initial findings point to a lack of mentoring, resources and funding as key issues young scientists face across the continent. Using this data, we will be able to identify critical areas in which young scientists need support and develop innovative strategies to alleviate these challenges.

The project comes at an important time as, over the past few years, African countries have initiated programmes to increase the number of PhD graduates. But if governments don't simultaneously develop support structures for graduates, and increase access to critical teaching and research infrastructure, these young scientists are set up to fail.

The study

The Global State of Young Scientists Africa project, uses an online survey (which is currently open to respondents) and in-depth interviews to gather as much detail as possible. It looks at young scientists' motivations, career ambitions and the barriers they experience in fulfilling their career aspirations.

It targets researchers and scholars who have earned a Masters or PhD within the last 10 years, irrespective of their current employment status and sector. It's also open to current PhD students in Africa and African scientists and scholars currently living in the diaspora.

Having this wide range of participants means the data will reflect a broad range of experiences. From early-career researchers with a history of moving within and out of Africa, to those who have never left their home countries. From department heads, to researchers who have trouble finding work despite their high qualifications. The team is also particularly interested in hearing from early career researchers outside of academia, as this helps us understand their reasons for not pursuing a career in research.

From our preliminary survey results – drawn from more than 700 young scientists' responses – we have found that, even with diverse backgrounds, early-career researchers have a great deal in common. A lack of mentoring, infrastructure, resources (staff and material) and funding for research and resources are key reasons for not pursuing a career in academia. There is also a strong desire for more training in grant writing and professional skills.

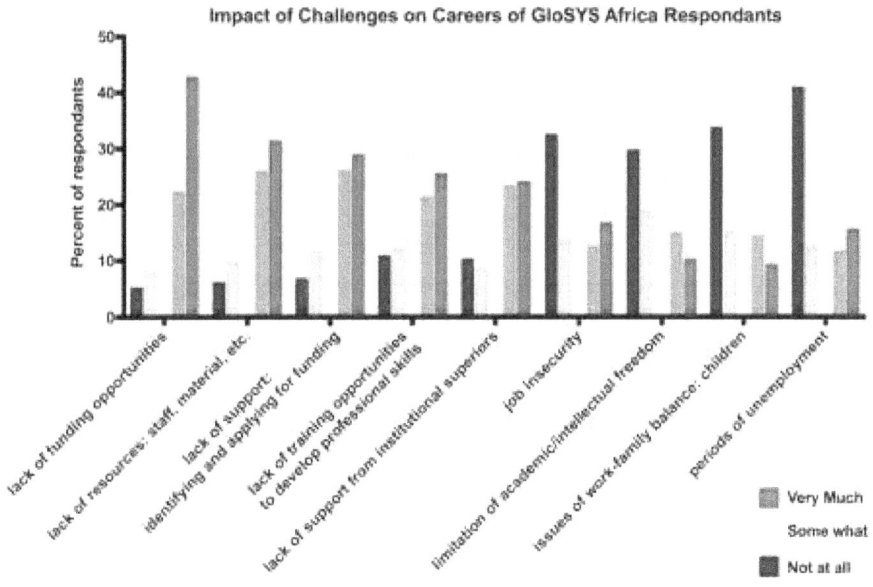

Using this information, the GYA plans to develop programmes to address the challenges, as we've previously done.

This is the third survey done under the Global State of Young Scientists umbrella. The first was a global study of young scientists from 14 countries across five continents. The second was a regional study which focused on four Southeast Asian countries.

A major challenge identified from those two studies was the desire for training in leadership skills. As these young scientists began to grow their own research groups they needed the tools to deal with the challenges of integrating research, teaching, and fundraising. In response, Global Young Academy members developed and implemented science leadership programmes in Africa and Asia, in collaboration with creative facilitators KnowInnovation and Future Africa.

Obtaining these new skills created an incentive for the young scientists to pursue their career in academia. The fellows found, for instance, the science leadership programmes to be one to the most significant workshops of their careers.

From barriers to action

The African leg of the survey continues. Once common challenges have been identified, the team will then work with policymakers in Africa as well as with international funding bodies to develop evidence-based initiatives to address them.

It's hoped that the Global State of Young Scientists Africa project will highlight further areas of need, so that the Academy can develop new innovative programmes in collaboration with science and education policymakers to improve young African scientists' prospects.

Early career researchers from and in Africa can become involved by taking the GloSYS Africa survey. The survey will remain open until mid September 2018 with results to be published at the beginning of 2019.

****

Black People of Africa and the African Diaspora Helping Each Other

AFRICAN SCIENTIFIC INSTITUTE

The Afrian Scientific Institute (ASI) has a network of Black notable scientists and Technologists from 64 countries. This expansive network serves as a "thinktank in cyberpace".

ASI has worked among Black scientists since 1967. Historically, ASI has solicited the presence of ASI Fellows at several of its activities, including our Science and Technology Awareness Fairs (1990 – 2000), which brought 3,500 to 4,000 children together each year for ten years to meet accomplished scientists and technologists from various companies, agencies, and academic institutions. We have also used the expertise of ASI Fellows during our Nursing Careers Outreach Program (2001 – 2003). We have used the images and accomplishments of ASI Fellows in our Blacks In Science Calendars (1989 – 1997), which was distributed nationally. We have used ASI Fellows to write articles for our Technology Transfer magazine (1979 – 1982) and SciTech Newspaper (1986 – 1991), each of which was distributed nationally. Since 2000, we have presented "Bridging the Divide" Mixers throughout the USA; analyzing facilities in Africa for rehabilitation; sponsoring a conference about supporting energy requirements in Africa; co-authoring Africa's next 50 year Manifesto; co-authoring "Civil Society Declaration" for signature by 65 countries' heads of state (including 15 Presidents) throughout Africa and the African Diaspora; and lecturing about advancements in science and technology.

The African Scientific Institute (ASI) has worked more than 50 years to motivate our youth to pursue careers in science and technology, while simultaneously enhancing the development of a worldwide network of professional scientists and technologists. ASI was founded in 1967. It is a non-profit organization representing a network of scientists, engineers, technologists, and health professionals, as well as young people aspiring to enter the world of science and technology.

Throughout ASI's history, it has always networked with professionals to enhance the capacity of scientists and technologists in our communities, within the U.S., Africa and the African Diaspora. To this end, ASI has extended its reach throughout international communities, wherever we might find scientists and technologists, to network our resources help solve challenges.

ASI has developed a vast network of international professionals, especially throughout Africa and the African Diaspora. Its network reaches approximately 15,000 professionals, of whom one half of them are scientists and technologists.

ASI has expanded its efforts to excite people about science and technology through our ASI Fellows Program, an international honorary association of accomplished individuals in the areas of "hard" science and technology. This collective of noteworthy individuals are to be found in academia, business and research at institutions and organizations on all over the world.

****

Conference Program
Monday, September 28, 2020

The Association of African Universities (AAU), in partnership with the African Union (AU) and the Government of Ghana, is inviting all educational stakeholders to the first Virtual African Academic Diaspora Homecoming, scheduled to take place from 28th -30th September 2020. The overall objective of the Homecoming is to strengthen collaboration and partnership among research and educational institutions of higher learning in Africa and the Diaspora to advance quality of education for all persons of African descent. The event is intended to be a major gathering of academics, policy makers, professional associations, research institutions, regional diplomatic missions and student associations from Africa and the diaspora. The event has four themes: (1) Role of the Diaspora in Higher Education and Innovation in Africa, (2) Strengthening Relationships between HBCUs and African Universities, and (3) Women of African Descent in Higher Education: Opportunities and Challenges, and (4) Higher Education: A Panacea to Racism, Equity and Promoting Social Justice.

****

Brazil Expanding Links in Africa: Lula's Positive Legacy

J.Peter Pham, PhD, October 12, 2010

The 2002 election of the former trade unionist from the left-wing *Partido dos Trabalhadores* (PT, "Workers' Party") caused a panic in the financial markets. *The Economist* last week highlighted the remarkable change.
.....................the part of Lula's foreign policy that deserves to be trumpeted is one that is largely overlooked: his leading his country into an unprecedented engagement with Africa.
Coming into office with the PT's left-wing bias in favor of South-South relations—one of his first acts of office was a federal decree that made the study of African history and African and Afro-Brazilian culture mandatory at all levels of Brazil's national curriculum—Lula reversed the preceding Fernando Henrique Cardoso administration's policy of closing diplomatic posts in Africa. During the first Lula term, Brazil not only reopened six shuttered embassies, but also opened thirteen new embassies and a consulate general so that the country now has a network of thirty-two embassies and two consulates general across the African continent. Clearly distinguishing itself from earlier Brazilian governments which focused their diplomatic attention on cultivating relations with the five Lusophone African countries—Angola, Cape Verde, Guinea-Bissau, Mozambique, and São Tomé and Príncipe—Lula's administration has broadened the scope of action for Brazilian diplomacy in Africa. The proof that Brazilian-African relations have intensified is that there are currently twenty-six embassies from African states resident in Brasilia and four African consulates general open in other Brazilian cities. During the past eight

years, Brazil has signed nearly two hundred bilateral accords with various African countries.

Lula is clearly personally invested in the building of ties Africa. During his two terms, he visited the continent on a dozen different occasions—most recently in July when he swung through Cape Verde, Equatorial Guinea, Kenya, Tanzania, and Zambia, en route to the Fifa World Cup final in South Africa. However, at the summit meeting of the Economic Community of West African States (ECOWAS) in Santa Maria, Cape Verde, he told the African leaders that "*Brazil—not just me—took a political decision to make a reencounter with the African continent*," explaining that Brazil could never repay its "*historical debt*" to Africa and that it "*would not be the country it is today with the participation of millions of Africans who helped build our country.*" Reviewing the trade, investment, and technology transfers that occurred under his administration, he laid a charge for his successor: "*Whoever comes after me has the moral, political, and ethical obligation to do much more.*"

Beyond the rhetoric of historical and political solidarity, however, there is substantive economic diplomacy with Africa, following upon the project for the creation of a free trade area initialed in 2000 between Brazil and the other member states of the South American Common Market (Mercosur)—Argentina, Paraguay, and Uruguay—and South Africa.

After having forgiven the relatively minor debts of Cape Verde ($2.7 million) and Gabon ($36 million) early in his presidency, Lula wiped out 95 percent of the public debt Mozambique owed to Brazil, some $315 million, in August 2004, and renegotiated the outstanding balance of $16 million. Subsequently, the Brazilian mining giant Companhia Vale do Rio Doce (CVRD, now Vale) was awarded a 25-year renewable contract to develop Mozambique's Moatize greenfields coal project that is expected to produce approximately 11 million tons of both metallurgical coal and thermal coal to be exported not only to Brazil, but also to Europe, Asia, and the Middle East. This project alone requires the participation of twenty other Brazilian firms, in addition to Vale, and will turn Mozambique into Africa's second largest coal producer, after South Africa. Its implementation will create 3,000 local jobs and the eventual production will employ 1,500 permanently.

Mozambique is not Vale's only investment in Africa. In March 2009, the company announced the creation of a join venture with the South African mining company African Rainbow Minerals Limited (ARM) aimed at enlarging the strategic options for growth in the

African copper belt. In addition, the company is also present in Angola, Guinea, South Africa, and the Democratic Republic of the Congo (DRC).

In Angola, the Lula administration extended lines of credit totaling $580 million in 2005, which enabled the Brazilian engineering and construction firm Construtora Norberto Odebrecht (CNO) to rebuild the war-damaged Capanda hydroelectric power plant. Subsequently, other subsidiaries of CNO's Salvador da Bahia-based parent company, Brazil Odebrecht, got involved in joint ventures with Angola's state-owned companies in diamonds and bio-fuels as well as commercial and residential real estate. Oldebrecht is now the largest private sector employer in Angola. Additional credits were subsequently extended to totals approaching $2 billion in conjunction with semi-public Petróleo Brasileiro S.A. (Petrobras) acquiring stakes in several offshore blocks in joint venture with the state-owned Sociedade Nacional de Combustíveis de Angola (Sonangol).

Angolan President José Eduardo dos Santos came away from his state visit to Brazil this past June with an additional $1 billion line of credit from his Brazilian counterpart to help fund reconstruction efforts in the African country after more than a quarter-century of war. On top of the $1.6 billion the Brasilia had previously committed to help Brazilian firms win those building contracts. The additional money ensures that Brazil will advance the lead it already enjoys as the leading funder of Angolan reconstruction—only China comes anywhere close to the capital for infrastructure provided by Brazil. From the strategic perspective, the money is being advanced not just to gain construction projects for Brazilian firms, as lucrative as these may be, but also because of a tacit understanding between the two Lusophone states that Petrobras will have preferential access to Sonangol's potential deposits offshore. The former's management is convinced that geological formations off of Angola are remarkably similar to Brazil's own pre-salt deposits which have already transformed the Latin America country into a rising oil power.

Not surprisingly, Brazil is seeking to replicate this model of generous investments in infrastructure in exchange for access to natural resources. In May, Brazil's ambassador to Tanzania, pledged to work closely with the nascent East Africa Community (EAC) of Burundi, Kenya, Rwanda, Tanzania, and Uganda. Meeting with EAC officials in the Kenyan resort of Naivasha, <u>Brazilian envoy Francisco Carlos Soares Luz</u> highlighted technology, energy, railways, environment, construction, and agriculture as among the sectors where his country might possibly cooperate with the regional bloc. The EAC is primed for engagement, all five of its member states having ratified this year the intergovernmental body's Common Market Policy.

During the Lula administration, Brazil's annual trade with African countries has quadrupled in value from $6 billion in 2003 to roughly $25 billion today. These figures represent an extraordinary increase of exports by an average of 28 percent per year and imports from Africa of about 23 percent per year. In terms of total volume of bilateral trade, Africa is taken as a whole ranks fourth among Brazil's top partners, ranking behind only the United States, China, and Argentina.

In August, Sudanese companies signed $500 million worth of deals with Brazilian engineering and construction firms following a visit by a Sudanese government delegation to Brazil. Significant among the accords are two deals involving the Kenana Sugar Company, Sudan's largest, whereby Brazil's Dedini is to provide machinery and equipment to double the size of Kenana's ethanol plant in Sudan and to set up a new biodiesel operation. Sugar is a key commodity in Sudan, where the population is sensitive to price hikes. The country, which hopes to be a sugar exporter by 2014 thanks to the deal, currently has to import to cover domestic consumption of 1.2 million tons a year.

The month, state-controlled Banco do Brasil joined forces with private competitor Bradesco to found a partnership with Portugal's Banco Espiritu Santo to expand activities in Africa. Initially the joint venture will target opportunities in Algeria, Angola, Cape Verde, Mozambique, Morocco, and South Africa.

Not surprisingly, perhaps nowhere on the continent is Lula's diplomacy more appreciated in Africa than in South Africa, one of Brazil's most important African trade partners. A recent editorial in the *Mail and Guardian* noted that *"South Africa has been the major beneficiary of Lula's South-South diplomacy, which helped Africa's most prosperous nation to strengthen its international standing and strategic weight"* and expressed concern that his likely successor, Rousseff, is *"an uninspiring technocrat [who] is unlikely to focus on Africa as much and may neglect it altogether as she faces formidable domestic challenges, such as urgent tax and pension reform."*

Lula has also been a driving force behind a loose political alliance of India, Brazil, and South Africa, formally called the "India-Brazil-South Africa (IBSA) Dialogue Forum," which was launched in 2004 with the goal of achieving common positions at the UN, the Doha Rounds, and other multilateral settings for the three major "southern" nations. Formal summits of the leaders of the IBSA states have so far been held in Brasilia (2006), Pretoria (2007), New Delhi (2008), and again in Brasilia in April of this year. The most recent meeting also featured at its margins an IBSA academic forum, a forum for parliamentarians, a round table on local governance, a women's forum, a forum for editors, and a networking session for CEOs from the member states.

It is important to note that Brazil's network of relations throughout African nowadays not only extends beyond the longstanding affinity for the continent's Lusophone countries on the part of the Itamaraty (as its Ministry of External Relations is often known from the palace that housed it in Rio de Janeiro and an eponymous edifice in Brasilia) to commercial and broader geopolitical interests, but it embraces a host of programs in the social and humanitarian sectors. In 2007, for example, Brazil provided the government of Ghana with technical assistance in designing of a pilot welfare program called the "Livelihood Empowerment Against Poverty" (LEAP). Experts from Brazil's Ministry of Social Development (MDS) took part in three missions to Ghana share knowledge about Brazil's *Bolsa Família* ("Family Allowance," a signature initiative of the Lula administration that gives financial aid to poor families in exchange for having their children vaccinated and keeping them in school), *Cadastro Único para Programas Sociais*(*CadÚnico*, "Single Registry for Social Programs," a program began under Lula's predecessor, but expanded broadly during his administration), and child labor eradication program (PETI). Due to the success of this exchange, the following year MDS partnered with the United Kingdom's Department for International Development (DFID) and the United Nations Development Programme (UNDP) International Poverty Center to launch the Africa-Brazil Cooperation Program on Social Development to promote international technical cooperation between developing countries to foster social protection programs. So far, programs have included technical assistance from Brazilian government representatives for African government agencies, study tours of Brazil for African officials, and internet-based distance learning opportunities.

Agriculture is another area that Brazil is uniquely positioned to help African countries and, in fact, has already done so. In a little more than generation, Brazil's own agricultural sector has gone from virtually anemic levels to a $250 billion per year concern that accounts for 35 percent by a combination of targeted investments in research to increase yields and financing guarantees to encourage the adoption of new technologies and equipment. In fact, *Empresa Brasileira de Pesquisa Agropecuária* (EMBRAPA, the "Brazilian Enterprise for Agricultural Research"), the affiliate of the Ministry of Agriculture whose pure and applied scientific work, diffused through a network of more than forty research and extension centers throughout Brazil, is credited with facilitating the country's agricultural leap, opened an office in Accra, Ghana, in 2006. From there EMBRAPA is spreading its know-how across Africa in 35 projects ranging from bolstering Angola's National Research Institute to helping Senegal develop its rice sector to assisting Tanzania with developing its dairy industry. Significantly, EMBRAPA officials have been careful to coordinate and align their programming with those specialized agencies of the African Union such as the Comprehensive African Agricultural Development Programme (CAADP) of the New Partnership for Africa's Development (NEPAD) as well as those of the continent's subregional organizations like the Southern African Development Community (SADC) Food, Agriculture and Natural Resources (FANR) Program and the West and Central African Council for Agricultural Research and Development (WECARD).

In addition to spearheading the UN operation in Haiti, currently the organization's third-largest, Brazil also currently contributes military and police personnel to four blue-helmeted peacekeeping missions in Africa: the UN Mission for the Referendum in Western Sahara (MINURSO), the UN Mission in Liberia (UNMIL), the UN Mission in Sudan (UNMIL), and the UN Operation in Côte d'Ivoire (UNOCI). As one Brazilian political scientist noted in a report in the Economist last month, there is a calculated strategy behind this participation: "*Brazil's elite thinks peacekeeping is part of the price you have to pay to be among the nations who make the rules.*" To this end, in 2005 the Lula administration created a peacekeeping school, the *Centro de Instrução de Operações de Paz* (CIOpPaz), near Rio de Janeiro. Since it opened its doors, CIOpPaz—which earlier this year changed its name to the Centro Conjunto de Operações de Paz do Brasil (CCOPAB, "Joint Center for Peacekeeping Operations of Brazil")—has trained more than 15,000 Brazilian army, navy, and air force personnel as well as police and civilians for deployment on UN peacekeeping missions.

Whilethe growth in influence of any other significant international actor on an African continent that increasingly holds geostrategic importance for interests of the United States is a development that bears very careful watching, there are a number of reasons why Brazil's increased engagement in Africa, unlike that of some others, ought to be cautiously welcomed.

First, there is no doubt that Africa stands to benefit from the addition of Brazil to the list of countries seeking access to the continent's natural resources and markets as well as political and strategic partnerships with African states. This is especially true if African leaders are able to develop a strategic approach that leverages their strengthened bargaining position.

Second, Brazil's *modus operandi* on the continent not only benefits the Latin American country, but it also carries a significant upside for Africans. Unlike many other "new actors" in Africa which, more often than not appear more as predators interested only in extracting commodities, Brazil has encouraged technology transfers to its African partners and often, as in the case of its efforts to spread the production of ethanol, endeavors to create a global market that will generate profits for its businesses as well as those of others.

Third, the lessons to be learned from Brazil's steady economic growth, agricultural success, and democratic politics are ones that African countries would do well to study closely. While one should not overlook Brazil's weaknesses, it should be acknowledged that the country has free and vigorous press that regularly uncovers and reports on these shortcomings. In this respect, Brazil is more like India than China or Russia among the BRIC (Brazil, Russia, India, and China) countries.

Fourth, the burgeoning Brazilian strategic engagement with Africa augurs for security and stability in Africa, especially if Lula's successors can be persuaded to not only strengthen their commitment to peacekeeping on the continent, but also to specifically cultivate military ties with Brazil's partners there, especially among the Lusophone countries. Quite frankly, the influence of the United States and other Western nations in some of these countries is somewhat limited either for historical reasons (e.g., backing the losing party during the long civil war in Angola) or simply lack of engagement (e.g., Guinea-Bissau, where the U.S. Embassy has been closed since 1998 despite the fact the country has become a major drug transshipment hub). Brazil, in contrast, is well-positioned to engage in confidence building and capacity strengthening in the security sectors of these countries.

Fifth, while Brazil will, of course, pursue its own objectives, not those of another country, in its diplomacy, there are considerable common interests such that, on the whole, Brasilia's impact ought to be viewed as positive. Notwithstanding his occasional diplomatic naïveté and his weakness for the left wing, Washington in particular ought to prefer Lula's more substantive investment in Africa to the attempts by Venezuela's Hugo Chávez to buy adhesion to his anti-Western transatlantic alliance. Moreover, tensions will diminish since Lula's likely successor, Dilma Rousseff, will likely focus on the domestic economic and social issues are her strength as thus adopt a less abrasive international style. Consequently, steps might be taken by both the government and the private sector to enhance the U.S.-Brazilian relationship overall and foster cooperation in Africa that advances both countries' interests in promoting good governance, supporting economic growth and development, increasing access to health and educational resources, and helping to prevent, mitigate, and resolve conflicts on the continent.

American and other policymakers, analysts, and businesses cannot afford to ignore the greater role that Brazil under Lula has successfully staked out for itself as a political and economic force on in Africa. Hence, as the United States continues advancing its own strategic interests on the continent as well as promoting those of African nations peoples, an active engagement by the North American superpower with the South America giant suggests itself as the policy option to be privileged.

FamilySecurityMatters.org Contributor J. Peter Pham is Senior Vice President of the National Committee on American Foreign Policy in New York City. He also hold academic appointments as Associate Professor of Justice Studies, Political Science, and African Studies at James Madison University in Harrisonburg, Virginia, and non-resident Senior Fellow at the Foundation for the Defense of Democracies in Washington, D.C. He currently serves as Vice President of the Association for the Study of the Middle East and Africa (ASMEA) and Editor-in-Chief of its refereed Journal of the Middle East and Africa.

Dr. Pham has authored, edited, or translated over a dozen books and is the author of over three hundred essays and reviews on a wide variety of subjects in scholarly and opinion journals on both sides of the Atlantic. In addition to the study of terrorism and political violence, his research interests lie at the intersection of international relations, international law, political theory, and ethics, with particular concentrations on the implications for United States foreign policy and African states as well as religion and global politics.

****

from **Science and Development Network (SciDev/Net)**
Africa and Brazil to cross-fertilise agricultural ideas
Busani Bafana, 27 July 2010

[OUAGADOUGOU, BURKINA FASO] An ambitious development partnership aimed at strengthening agricultural collaboration between Africa and Brazil was launched at the 5th African Agriculture Science Week in Burkina Faso last week (21 July).

The initiative — the Africa-Brazil Agriculture Innovation Marketplace — recently announced in Brazil, and now formally launched, aims to enhance South–South knowledge and technology transfer and stimulate policy dialogue between the two regions.

It will also promote collaborative research projects, according to Paulo Duarte, technical coordinator at Embrapa, the Brazilian Agricultural Research Corporation. He told *SciDev.Net* that funding for joint project proposals with Embrapa for the coming two years will be approved at a meeting in Brazil in October. Up to seven projects will be selected from more than 40 proposals already submitted.

"Innovation cannot happen without development," said Monty Jones, co-chair of the new marketplace and executive director of the Forum for Agriculture Research in Africa (FARA), which signed a memorandum of understanding with Embrapa, formalising the marketplace launch at the event in Burkina Faso.

Africa has the potential to be the world's food basket, said Jones, but poor technology and innovation adoption remain a constraint. This would be greatly improved with better exchange of science and technology knowledge, he said.

Jones praised the Africa–Brazil initiative as a major step towards improving knowledge and innovation to boost African agriculture. He said Africa had developed some promising technologies, such as banana tissue culture and New Rice for Africa, but too many other technologies remained on the shelf or had made little impact because of poor dissemination of information.

"Brazil has come a long way to feed its people because of an innovative approach to science and technology," said Duarte. "Working with Brazil, Africa should not reinvent the wheel but adopt some of these methodologies to catalyse its agricultural development."

In May this year, Brazil invited ministers and officials from more than 30 African countries to a meeting, 'Brazil-Africa Dialogue on Food Security, Combating Hunger and Rural Development', organised by the Brazilian Ministry of Foreign Relations, where the marketplace was first announced.

****

CUBA-AFRICA: Decades of Assistance and Cooperation
Inter Press Service - July 2, 2004 **Patricia Grogg**

Dozens of countries in Africa benefit from medical or other assistance programmes from Cuba, and in four decades this socialist island nation has helped train around 30,000 young people from that continent in a number of specialties.

Medical assistance, which began in 1963 when Cuba sent health brigades to Algeria, has been extended to more than 20 nations, 16 of which are included in the Integral Health Programme, which also encompasses seven countries in Latin America and two in Asia.

"Cooperation in that field is still the most important. By the end of 2003, we had 2,574 collaborators in Africa, most of them in the area of health," Milagros Franco, the Ministry of Foreign Investment's director for Africa, told IPS.

The aid has had a strong impact. According to official statistics, infant mortality in areas where Cuban medical professionals provide assistance has plunged: from 59 to 7.8 per 1,000 live births in Ghana, from 48 to 10.6 in Eritrea, and from 131 to 35.5 in Equatorial Guinea.

The assistance includes training of health workers in the areas where Cuban doctors provide their services and of health professionals in institutes in Cuba.

Havana has also helped set up medical schools in Gambia and Equatorial Guinea, and Cuban professors teach in medical schools in several countries, including Ethiopia, Uganda and South Africa.

Cuba has diplomatic ties with 53 African nations and provides assistance to 51. Although the greatest emphasis is put on health and education, cooperation also includes the areas of sports, construction, agriculture and urban planning.

"The cooperation is based in first place on the training and development of human resources and on our commitment to that region, with which we have historical, as well as blood, ties," said Franco.

According to official statistics, around 6.3 percent of Cuba's 11.2 million people are university graduates, and Cuba has around 64 centres of higher learning, in which a combined total of 200,000 students are enrolled.

Political leaders in Africa especially value the aid provided by Cuban doctors in fighting pandemics like tuberculosis and HIV/AIDS, in clinics which are often located in remote regions, and in specific programmes in the area of prevention, for instance.

As part of an agreement signed with Botswana in 2002, Cuban health professionals work in several hospitals and clinics that specialise in treating those living with HIV/AIDS in that southern African nation.

An estimated 29 million people are living with HIV/AIDS in sub-Saharan Africa.

Cuba helped many countries in that region in their independence struggles, but the fight has not ended -- it is just a different war now, a government official from Botswana recently told the press on a visit to Havana.

Botswana, which has a population of 1.8 million, has the highest proportion of people living with HIV on the planet, with 36 percent of people between the ages of 15 and 49 infected.

"All of this collaboration is carried out on a basis of respect, with the idea of resolving concrete problems, and without taking jobs from professionals from those countries," said Liber Puente, a specialist in relations with sub-Saharan Africa in Cuba's Foreign Ministry.

Puente added that many youngsters who have studied on scholarships from the Cuban government are today professionals working for the development of their countries and holding political leadership and government posts.

Claude Grace Uushona, Namibian Ambassador in Havana, is one such case. She first arrived in Cuba at the age of 16, injured and with horror written on her face after the massacre she witnessed in a refugee camp in Cassinga, in southern Angola.

"It was a camp for women and children refugees located 250 km from the border with Namibia. On May 4, 1978, South African forces bombed us with military helicopters and planes," she told IPS.

The survivors, many of whom were pre-adolescents and adolescents, were invited by President Fidel Castro to travel to Cuba to attend school on the Isla de la Juventud (the Island of Youth), to the south of Cuba.

"There were about 600 of us who came. I went to high school and prep school there. Then I studied at the University of Gambia and I became the first woman governor in Namibia. Now I am very happy to be in my second country again," Uushona added.

Cuba and Namibia established diplomatic ties on Mar. 21, 1990, the day the southwest African nation declared independence. Since then the two countries have engaged in a number of cooperation programmes in areas like health, education and sports.

Namibia is now the biggest recipient of Cuban investment in Africa, with two Cuban-Namibian fishing companies and mixed ventures that manufacture guayaberas -- the typical Caribbean men's shirt -- and generic medicines.

Uushona said the companies are very important for the development of Namibia.

The ambassador recalled that the victory of Cuban and Angolan troops fighting alongside the guerrilla Southwest African People's Organisation (SWAPO) against South Africa's apartheid army in the April 1988 battle of Cuito Cuanavale in southern Angola played a decisive role in the liberation of Namibia.

Pretoria, after it was defeated in its last major offensive in Angolan territory, was forced to accept the tripartite accords with Angola and Cuba that led to the withdrawal of South Africa's army from Namibia, and in November 1989 Namibia held elections for a constituent assembly.

Since then, SWAPO has been the governing party in Namibia.

The tripartite accords also led to the gradual withdrawal of the Cuban military contingents from Angola, and the start of the abolition of South Africa's racist apartheid regime.

An estimated 300,000 Cubans passed through Angola between 1975 and 1989, including 2,016 who were killed. (END/IPS/LA CA/HE IF IP DV/TRASP-SW/PG/MP/04)

Conclusion

While we are all excited about past and present contributions by Black people to the world of science and technology, we must focus our attention to the future development of Africa and Black people throughout the African Diaspora.

Examples of our achievements are essential to help build our confidence to engage necessary actions of our future. When we have courage to independently move forward, then we gain respect as an equal partner and are offered equal access to collaborative work.

Bibliography

1. My book *Spirits in the Whirlwind*, the 7 keys: RISEPEM.
2. *African Glory* (the story of vanished Negro civilizations), by J.C. deGraft Johnson, with Introduction by John Henrik Clarke
3. *Blacks in Science and Technology*, *Black Achievers in Science and Technology* (a directory of ASI Fellows), both by the African Scientific Institute
4. *The African Past*, by Basil Davidson
5. *From Slavery to Freedom* (a history of African Americans), by John Hope Franklin
6. *West African Resistance* (the military response to colonial occupation), edited by Michael Crowder
7. 11 Volume *The History of Civilization*, by Will and Ariel Durant (at least volumes 1, 2, 3)
8. *The Mis-Education of the Negro*, by Carter G. Woodson
9. *The Destruction of Black Civilization*, *by* Chancellor Williams
10. *Neo-African literature* (a history of Black writing), by Janheinz Jahn
11. *Golden Age of the Moor*, by Ivan van Sertima
12. *They Came Before Columbus*, by Ivan van Sertima
13. *Blacks in Science and Technology* (ancient and modern), by Ivan van Sertima
14. *The African Origin of Civilization*, by Cheikh Anta Diop (a physicist from Senegal)
15. *What They never Told You in History Class*, by Indus Khamit-Kush
16. *Neo-Colonialism* (the last stage of imperialism), by Kwame Nkrumah
17. *Consciencism*, by Kwame Nkrumah
18. *African Achievements* (leaders, civilizations and cultures of ancient Africa), by Lester Brooks
20. *The Wretched of the Earth*, by Frantz Fanon (a scientist/psychiatrist from Martinique in 1925)

21. *Toward the African Revolution* (essays, notes, letters), by Frantz Fanon

22. *Developing Positive Self-Images & Discipline in Black Children*, by Jawanza Kunjusfu

23. *African Intellectual Heritage* (a book of sources), by Molefi Kete Asante and Abu S. Abarry, editors

24. *Dictionary of Black African Civilization*, by Georges Balandier, Jacques Maquet, et al

25. *African Americans in Science, Math, and Invention*, by Ray Spangenburg and Kit Moser

26. *African American Firsts in Science & Technology*, by Raymond Webster (an ASI Fellow who has a database of thousands of patents by Black people)

27. *USA* **QuickFacts** from the *US* Census Bureau: In 2010 there are about 40,000,000 Black people in the US. Per the **Ontario Black History Society**: The total population of all the people in Canada as of January 2007 is an estimated 33,777,304. Of this number, an estimated 662,200 identified themselves as Black in the year 2001.

28. Per answers from various sources, there seems to be about 8%-10% Black people in France, 4 – 5,000,000 and about 2,000,000 people of African descent living in England. The rest of Europe has about 1,000,000 (primarily Germany, Spain and Portugal)

29. Black people in China:

30. Black people in Indonesia:

31. *Indigenous People of Africa and America Magazine*

32. *Wikipedia*

33. Compiled by Clifton L. Holland, Director of Latin American Socio-Religious Studies Program_Programa Latinoamericano de Estudios Sociorreligiosos (PROLADES), (last revised on July 30, 2005)

34. *Black Inventors*, by Keith Holmes

35. See **Appendix:** *"Africa and Brazil to cross-fertilise agricultural ideas". "Brazil Expanding Links in Africa: Lula's Positive Legacy".*

36. See Appendix: "CUBA-AFRICA: Decades of Assistance and Cooperation".

37. See Appendix: *"U.S. PATENTS (few of the thousands of patents by Black people)"*

38. *The Inventive Spirit of African American*, by *Patricia* Carter *Sluby*

Other Bibliographic sources of information about African American Contributions to Science and Technology not included in this paper include:

- Allen, Sallie T. The Directory of Black Nursing Faculty; Baccalaureate and Higher Degree Programs. Lisle, IL: Tucker Publications, Inc.,1988. (R, 610.7302573, qA429, 89-2808).
- American Nurses' Association. Minority Groups in Nursing: A Bibliography. Kansas City, MO: The Association, 1976. (C, 016.3621, A512, 76-8196, 1976).
- Baker, Henry Edwin. The Colored Inventor. Reprint of the 1913 edition. New York: Arno Press, 1969. (C, 608.717496, B167, 78-21142). Originally published on the 50th anniversary of the Emancipation Proclamation.
- Baker, Henry Edwin. "The Negro as an Inventor." In: Twentieth Century Negro Literature, or, A Cyclopedia of Thought on the Vital Topics Relating to the American Negro, edited by D. W. Culp. Naperville, IL: J. L. Nichols & Co., 1902. (C, 325.26, C96). Contains a list of several hundred patents issued to African-Americans through the year 1900, a commentary on some of the inventors, a summary of background research, and a brief biography of the author.
- Banneker, Benjamin. Works of Benjamin Banneker. New York: 3M International Microfilm Press, 1970. (MA/FM, 528.1, B219). Microfilmed from the Schomburg Collections of the New York Public Library. Contents: Benjamin Banneker's Pennsylvania, Delaware, Maryland and Virginia Almanack and Ephemeris for the Year of Our Lord, 1792; Banneker's Almanack and Ephemeris for the Year of Our Lord, 1793; Virginia Almanack, for the Year of Our Lord, 1794; John H. Latrobe, Memoir of Benjamin Banneker, read before the Maryland Historical Society; Martha Tyson Ellicot, Sketch of the Life of Benjamin Banneker, from Notes Taken in 1836; Benjamin J. Lossing, "Benjamin Banneker," In: Eminent Americans: Comprising Brief Biographies of Three Hundred and Thirty Distinguished Persons.
- Burt, McKinley, Jr. Black Inventors of America. Portland, OR: National Book Company, 1989. (C, 926.6048 qB973, 90-13120).

- Carnegie, Mary Elizabeth. The Path We Tread; Blacks in Nursing, 1854-1984. Philadelphia: Lippincott, 1986. (C, 331.4816107, C289, 87-053723).
- Carver, George Washington. George Washington Carver in His Own Words. Edited by Gary R. Kremer. Columbia: University of Missouri Press, 1987. (C, 630.924, C331, 87-33768).
- Cobb, William Montague. The First Negro Medical Society; A History of the Medico-Chirurigical Society of the District of Columbia, 1884-1939. Washington, DC: The Associated Publishers, 1939. (Z, 610.6, M486).

- Conner, Douglas L. A Black Physician's Story: Bringing Hope in Mississippi. Jackson, MS: University Press of Mississippi, 1985. (C, 610.924, C752, 88-028184).
- Dummett, Clifton Orrin, ed. The Growth and Development of the Negro in Dentistry in the United States. Chicago: National Dental Association, 1952. (Z, 617.6, D889).
- Greene, Harry Washington. Holders of Doctorates Among American Negroes: An Educational and Social Study of Negroes Who Have Earned Doctoral Degrees in Course, 1876-1943. Boston: Meador Publishing Company, 1946. (C, 325.26, G7992).
- Guzman, Jessie Parkhurst. George Washington Carver, A Classified Bibliography. Records and Research Pamphlet No. 3. Tuskegee, AL: Department of Records and Research, Tuskegee Institute, 1954. (C, 012, C331, 74-7625).
- Haber, Louis. Black Pioneers of Science and Invention. New York: Harcourt, Brace & World, 1970. (Z, 509.2, Ah12).
- Hardwick, Richard. Charles Richard Drew, Pioneer in Blood Research. New York: Scribner, 1967. (Z, 610.92, D776h2).
- Harris, M. A. Granville T. Woods Memorial. Collector's ed. Brooklyn: M.A. Harris, 1974. (C, 620.0092, qW894, 75-6844).
- Ho, James K. K., ed. Black Engineers in the United States; A Directory. Washington, DC: Howard University Press, 1974. (C, 620.002573, H678, 77-29269).
- "Inventors and Scientists," In: The Negro Almanac: A Reference Work on the Afro-American, edited by Harry A. Ploski and James Williams. 4th ed. New York: John Wiley & Sons, 1983, pp. 1053-1076. (R, 973, qP729a, 1983).
- Ives, Patricia Carter. Creativity and Inventions: The Genius of Afro-Americans and Women in the United States and their Patents.

Arlington, VA: Research Unlimited, 1987. (C, 609.22, I95, 90-14537).
- James, Portia. The Real McCoy: African-American Invention and Innovation, 1619-1930. Washington, DC: Smithsonian Institution, 1989. (C, 609.22, J28, 90-13178).
- Jenkins, Edward S., ed. American Black Scientists and Inventors. Washington, DC: National Science Teachers Association, 1975. Available as ERIC document. (ED 176-983).
- Kenney, John A. The Negro in Medicine. Tuskegee, AL: Tuskegee Institute Press, 1912. (Z, 610.92, Ak).
- Latimer, Lewis H. Incandescent Electric Lighting: A Practical Description of the Edison 621.326, L35).
- Manning, Kenneth R. Black Apollo of Science: The Life of Ernest Everett Just. New York: Oxford University Press, 1983. (C, 574.0924, M283, 84-29211).
- Morais, Herbert Montfort. History of the Negro in Medicine. International Library of Negro Life and History, vol. 4. New York: Publishers Company, Inc., 1968. (C, 917.3097496, qI615).
- Organ, Claude H., Jr. and Margaret M. Kosiba. A Century of Black Surgeons; the U.S.A. Experience. Norman, OK: Transcript Press, 1987. (C, 617.0922, c397, 88-033590).
- Pearson, Willie, Jr. and H. Kenneth Bechtel, eds. Blacks, Science and American Education. New Brunswick, NJ: Rutgers University Press, 1989. (C, 508.996073, B631, 89-45053).
- Sammons, Vivian O., comp. Blacks in Science and Related Disciplines. LC Science Tracer Bullet TB-85-5. Washington, DC: Library of Congress, Science Reference Section, June 1985. (LC 33.10:85-5).
- Blacks in Science and Medicine. New York: Hemisphere Publishing Corporation, 1990. (C, 509.2, qB631, 90-6929).
- "Scientists." In: Encyclopedia of Black America, edited by William A. Low and Virgil A. Clift. New York: McGraw-Hill Book Company, 1981. (R, 973.0496073, qE56, 81-24479).
- Stanley, Autumn. "From Africa to America: Black Women Inventors." In: The Technological Woman: Interfacing with Tomorrow, edited by Jan Zimmerman. New York: Praeger, 1983. (C, 305.4, T255, 83-29717).
- Summerville, James. Educating Black Doctors; A History of Meharry Medical College. University, AL: University of Alabama Press, 1983. (C, 610.7117685, S955, 84-37541).

- Williams, James C., comp. At Last Recognition in America: A Reference Handbook of Unknown Black Inventors and Their Contributions to America. Chicago: B.C.A. Publishing Corp., 1978. (C, 608.778, W724, 82-24801, v.1).

SELECTED PERIODICAL ARTICLES

- Baker, Henry E. "The Negro in the Field of Invention." Journal of Negro History, vol. 2, no. 1, Jan. 1917, pp. 21-36. (J, 325.26, J86).
- Beardsley, E. H. "Making Separate, Equal: Black Physicians and the Problems of Medical Segregation in the Pre-World War II South." Bulletin of the History of Medicine, vol. 57, no. 3, 1983, pp. 382-396. (J/FM, 610.9, B936).
- Christopher, Michael C. "Granville T. Woods; The Plight of a Black Inventor." Journal of Black Studies, vol. 11, no. 3, Mar. 1981, pp. 269-276. (J/FM, 310.45196, J862).
- Congressional Record. 53d Cong., 2d sess., vol. 26, pt. 8, Aug. 10, 1894, pp. 8382-8384. (J, 328.73, qU49cr). Contains "A Partial List of Patents Granted by the United States for Inventions by Afro-Americans," appended to the record by Congressman Murray of South Carolina.
- Drew, Charles Richard. "Negro Scholars in Scientific Research." Journal of Negro History, vol. 35, no. 2, Apr. 1950, pp. 135-149. (J, 325.26, J86). Address delivered by Dr. Drew before the Annual Meeting of the Association for the Study of Negro Life and History in New York City on October 30, 1949.
- Gray, Garry. "H. C. Haynes, Barber and Inventor." Negro History Bulletin, vol. 40, no. 5, Sept.-Oct. 1977, pp. 751-752. (J, 325.2670973, qN393).
- Ives, Patricia Carter. "Patent and Trademark Innovations of Black Americans and Women." Journal of the Patent Office Society, vol. 62, no. 2, Feb. 1980, pp. 108-126. (LAW/PER). Contains appendices with lists of patents awarded to blacks and women. Names many African-American patentees, especially in the 60s and 70s, not included in this bibliography.
- Jenkins, Edward S. "Impact of Social Conditions; A Study of the Works of American Black Scientists and Inventors." Journal of Black Studies, vol. 14, no. 4, June 1984, pp. 477-491. (J/FM, 310.45196, J862).
- Odom, Retha. "Pioneers in Technology." Black Enterprise, vol. 11, no. 11, June 1981, pp. 182-188. (J, 338.04, qB627, 79-55791).

- "Seminar Series Honors Blacks in Science and Technology." Science, vol. 216, Apr. 2, 1982, pp. 43-44. (J/FM, 505,S411).
- Young, Herman A. and Barbara H. Young. "Science and Black Studies." Journal of Negro Education, vol. 46. no. 4, 1977, pp. 380-387. (J/FM, 371.974, J86). Contains "Micro-Roster of Black Scientists and Inventors."

SPECIAL MATERIALS
- Akinsheye, Dexter. African-American Inventors Series. Silver Spring, MD: Three Dimensional Publishing, 1989. A collection of 39 study prints which includes patent diagrams, photographs and histories of some African-American inventors.
- Ford, C. W. (1994). We can all get along. New York: Dell.
- Malcom, S. M. (1990). Reclaiming our past. Journal of Negro Education, 59(3), 246-59.
- Pearson, W., Jr. (1989). The future of blacks in science: Summary and Recommendations. In W. Pearson, Jr. & H. K. Bechtel (Eds.), Blacks, science, and American education (pp. 137-152). New Brunswick, NJ: Rutgers University Press [ED 325 353]
- Pearson, W., Jr., & Bechtel, H. K. (Eds.). (1989). Blacks, science, and American education. New Brunswick, NJ: Rutgers University Press. [ED 325 353]

COLLECTIVE BIOGRAPHIES

- African-American scientists. McKissack, P., & McKissack, F. Brookfield, CT: Millbrook Press, 1994. 96 p. Q141.M36 1994 Examines the lives and achievements of African American scientists from colonial days to the present.
- Against all opposition: Black explorers in America. Haskins, J. New York: Walker, 1992. 86 p. Bibliography: p. 83-84.
- American Black scientists and inventors. Washington, DC: National Science Teachers Association, 1975. 79 p. Q141.A46. Contents: Jenkins, E.S. Ernest E. Just, cell physiologist; Hudson, G.H. Garrett Augustus Morgan: Big Chief Mason, ingenious American; Ryder, E.C. George Washington Carver, agricultural scientist; Jackson, W.S. Benjamin Banneker, Black astronomer; Jenkins, E.S. Percy L. Julian, soybean chemist; Jackson, W.S. Granville T. Woods, railway communications wizard; Hudson, G.H.

Charles Richard Drew, blood plasma pioneer; Hudson, G.H. Charles Henry Turner, scientist, teacher, author, humanitarian.--Hudson, G.H. Matthew A. Henson, famous explorer; Jenkins, E.S. Leon Roddy, spider man; Ryder, E.C. Elijah McCoy, inventor; and Ryder, E.C. Daniel Hale Williams, pioneer heart surgeon. Includes bibliographies.
- An African-American Bibliography: Science, medicine, and allied fields. Selected resources from the collections of the New York State Library. Strasser, T. C. Albany, NY: New York State Library, 1991 [ED 332 713]. This bibliography was issued in honor of both Black History Month and Inventors Day in February 1991. It focuses on the contributions of African Americans in the areas of science, technology, medicine, and allied fields such as dentistry and nursing. The materials cited emphasize the accomplishments of individuals from all parts of the United States, in all periods, and from all backgrounds. The bibliography includes books, selected periodical articles, patents, and other materials.
- Black Americans in aviation. Peters, R. E., & Arnold, C. M. San Diego, CA: Neyenesch Printers, 1975. 85 p. Bibliography: p. 83. Tl553.P47
- Blacks in science: Astrophysicist to zoologist. Carwell, H. Hicksville, NY: Exposition Press, 1977. 95 p. Bibliography: p. 93-94. Q141.C23
- Black inventors of America. Burt, M. Portland, OR: National Book Co., 1969. 143 p. T39.B87 1969
- Black mathematicians and their works. Edited by V. K. Newell and others. Ardmore, PA: Dorrance, 1980. 327 p. Includes Bibliographies. QA28.B58
- Black pioneers of science and invention Haber, L. San Diego: Harcourt Brace Jovanovich, 1991, 264 p. Bibliography: p. 245-254. Q141.H2 1991
- (Reprint. Originally published, New York, Harcourt, Brace & World, 1970.)
- Black scientists. Yount, L. New York: Facts on File, 1991. 111 p. Bibliography: p. 106. Q141.Y68 1991
- Created equal: the lives and ideas of Black American innovators. Brodie, J. M. New York: W. Morrow, 1993. 208 p. E185.96.B835 1993
- Creativity and inventions: The genius of Afro-Americans and women in the United States and their patents. Ives, P. C. Arlington,

VA: Research Unlimited, 1987. 88 p. Bibliography: p. 81-86. T21.184 1987
- 11 African-American doctors. Rev. and expanded ed. Hayden, R. C. Frederick, MD: Twenty-First Century Books, 1992. 206 p. R695.H39 1992
- The hidden contributors: Black scientists and inventors in America. Klein, A. E. Garden City, NY: Doubleday, 1971. 203 p. E185.8.K56
- 9 African-American inventors. Hayden, R. C. Frederick, MD: Twenty-First Century Books, 1992. 169 p. T39.H39 1992
- Outward dreams: Black inventors and their inventions. Haskins, J. New York: Walker, 1991. 101 p. Bibliography: p. 97-98. T39.H37 1991
- Profiles of pioneer women scientists. O'Hern, E. M. Washington, Acropolis Books, 1985. 264 p. Bibliography: p. 235-250. R153.O36 1985
- The real McCoy: African-American invention and innovation, 1619-1930. James, P. P. Washington, Published for the Anacostia Museum of the Smithsonian Institution by the Smithsonian Institution Press, 1989. 110 p. Bibliography: p. 103-106. T39.J28 1989
- A Salute to Black scientists and inventors. Chicago: Empak Enterprises, 1985. 28 p. (An Empak "Black history" publication series, v.2) Q141.S285 1985
- 7 African-American scientists. Rev. and expanded ed. Hayden, R. C. Frederick, MD: Twenty-First Century Books, 1992. 171 p. Q141.H37 1992

INDIVIDUAL BIOGRAPHIES
- Benjamin Banneker. Conley, K. New York: Chelsea House Publishers, 1989. 109 p. Bibliography: p. 106. QB36.B22C66 1989
- Chronicles the life of an 18th-century Black tobacco farmer who taught himself mathematics, astronomy, and clockmaking, became famous for his almanacs, and participated in the original survey of Washington, D.C.
- Charles Drew. Mahone-Lonesome, R.. New York: Chelsea House Publishers, 1990. 109 p. RD27.35.D74M34 1990
- A biography of the surgeon who conducted research on the properties and preservation of blood plasma and was a leader in establishing blood banks.
- Charles Richard Drew, M.D. Wolfe, R. New York: F. Watts, 1991. 64 p. Bibliography: p. 62. RD27.35.D74W65 1991

- A bibliography of the noted physician, focusing on his discovery of methods for separating plasma from blood.
- George Washington Carver. Gray, J. M. Englewood Cliffs, NJ: Silver Burdett Press, 1991. 138 p. Bibliography: p. 127-129. S417.C3G73 1990
- Describes the life and accomplishments of the former slave who became a scientist and devoted his career to helping the South improve its agriculture.
- Gifted hands. Carson, B., & Murphey, C. B. Washington, DC: Review and Herald Pub. Association, 1990. 232 p. RD592.9.C37A3 1990b
- The chief of pediatric neurosurgery at Johns Hopkins University hospital tells about his life and some of his cases.
- Lewis Howard Latimer. Turner, G. T. Englewood Cliffs, NJ: Silver Burdett Press, 1991. 128 p. Bibliography: p. 113-119. T40.L37T87 1990
- A biography of the African-American inventor who, among other contributions, invented an inexpensive method of manufacturing carbon filaments for electric light bulbs.
- The life of Charles Drew. Talmadge, K. S. Frederick, MD: Twenty-First Century Books, 1992. 84 p. Bibliography: p. 83. RD27.35D74T35 1991 A biography of the Black surgeon who was noted for his research on blood plasma.
- Man with a million ideas: Fred Jones, genius/inventor. Ott, V., & Swanson, G. B. Minneapolis: Lerner Publications Co., 1977. 109 p. T40.J59O87 1977 A biography of Frederick McKinley Jones, the Black engineer and inventor who is credited with many inventions, including refrigeration units for trucks and railroad cars, the portable x-ray unit, and the ticket dispenser.
- Mathematician and administrator, Shirley Mathis McBay. Verheyden-Hilliard, M. E. Bethesda, MD: Equity Institute, 1985. 31 p. QA29M38V47 1985 A brief biography of the woman mathematician who was the first African American to earn a Ph.D. from the University of Georgia.
- Matthew Henson. Gilman, M. New York: Chelsea House Publishers, 1988. 110 p. Bibliography: p. 108. G635.H4G55 1988 Follows the life of the Black explorer who accompanied Robert Peary on the expedition to the North Pole
- Paul Cuffe. Diamond, A. New York: Chelsea House Publishers, 1989. 111 p. Bibliography: p. 108. E185.97.C96D53 1989 A biography of the American seaman and merchant who encouraged

fellow Blacks to colonize Sierra Leone, sought a stronger legal position for Blacks in America, and was responsible for a Massachusetts law giving Blacks the right to vote.
- Queen Bess: daredevil aviator. Rich, D. L. Washington, DC: Smithsonian Institution Press, 1993. 153 p. Bibliography: p. 137-141. TL540.C646R52 1993 About Bessie Coleman.
- The real McCoy: the life of an African-American inventor. Towle, W., Paintings by Wil Clay. New York: Scholastic, 1993. 1 v. (unpaged). T40M43T68 1993 A biography of Elijah McCoy, the Canadian-born Black American who studied engineering in Scotland and patented over 50 inventions despite the obstacles he faced because of his race.
- Ronald McNair. Naden, C. J. New York: Chelsea House, 1990. 109 p. Bibliography: p. 106. TL789.85M36N34 1990 A biography of the Black astronaut who was a crew member aboard the ill-fated Challenger space shuttle mission that exploded on takeoff in January 1986.
- Shoes for everyone: a story about Jan Matzeliger. Mitchell, B. Minneapolis: Carolrhoda Books, 1986. 63 p. TS990. M335M58 1986 A biography of the half-Dutch/half-Black Surinamese man who, despite the hardships and prejudice he found in his new Massachusetts home, invented a shoe-lasting machine that revolutionized the shoe industry in the late 19th century.
- Scientist and administrator, Antoinette Rodez Schiesler. Verheyden-Hilliard, M. E. Bethesda, MD: Equity Institute, 1985. 31 p. QD22.S34V47 1985 Relates the story of an African American woman who overcame childhood difficulties with mathematics and went on to earn a Ph.D. in chemistry.
- Scientist and strategist, June Rooks. Verheyden-Hilliard, M. E. Bethesda, MD: Equity Institute, 1988. 31 p. Q143.R58V47 1988 A brief biography of June Rooks, a Black woman who contracted polio as a child, struggled against poverty, earned her degree in physics, and went on to become an operations research analyst with the U.S. Navy.
- Space challenger: the story of Guion Bluford; an authorized biography. Haskins, J., & Benson, K. Minneapolis: Carolrhoda Books, 1984. 64 p. Tl789.85B58H37 1984 Guy Bluford, the first Black American in space, was a crew member of the space shuttle Challenger on its August 1983 flight.

- Stop and go: Garrett Morgan, inventor. Sims, D. J. Los Angeles: Children's Cultu-Lit Book Co., 1980. 32 p. TE228S56 1980 A brief biography of the inventor of the traffic light and the gas mask.
- Sure hands, strong heart: the life of Daniel Hale Williams. Patterson, L. Nashville: Abingdon, 1981. 159 p. Bibliography: p. 157-159. RD27.35.W54P37 A biography of the Black surgeon who, among other achievements, was the first to perform open heart surgery.
- Trailblazer; Negro nurse in the American Red Cross. Pitrone, J. M. New York: Harcourt, Brace & World, 1969. 191 p. RT37.D3P5

- A biography of Frances Reed Elliott Davis whose determination to help relieve the physical pains of her people led her to become the first Black nurse enrolled by the American Red Cross.

BIOGRAPHICAL REFERENCE TOOLS

- A Salute to historic Black firsts. Publisher and editor, Richard L. Green. Chicago: Empak Pub. Co., c1989. 32 p. (An Empak "Black history" publication series, v. 7) E185.96S243 1989
- African American biographies: profiles of 558 current men and women. Hawkins, W. L. Jefferson, NC: McFarland, 1992. 490 p. E185.96.H38 1992.
- African American history: four centuries of Black life. Hughes, L., & Meltzer, M. New York: Scholastic, 1990. 312 p. E185.H83 1990 Revision of A pictorial history of Black Americans, 5th ed. (c1983)
- Afro-Bets Book of Black heroes from A to Z: an introduction to important Black achievers for young readers. Hudson, W., & V. Wilson Wesley. Orange, N.J., Just Us Books, 1988. 54 p. Bibliography: p. 52. E185.96.H77 1988
- Black history, Black lives: a comprehensive list of Black biographies for young people arranged by birthdate. Ginsberg, Dale Ann. Merion Station, PA: Anndale Books, 1986. 25 p. Z1361.N39G54 1986
- Blacks in science: Ancient and modern. Editor, Ivan Van Sertima. New Brunswick, USA., Transaction Books, 1983. 302 p. (Journal of African civilizations, v. 5, no. 1/2) DT14.J68, v. 5, no. 1/2 Bibliography: p. 295-297.

- Blacks in science and medicine. Sammons, V. O. New York: Hemisphere Pub. Corp., 1990. 293 p. Bibliography: P. 261-268 Q141.B58 1990
- The Black 100: a ranking of the most influential African-Americans, past and present. Salley, C. Seacaucus, NJ: Carol Pub. Group, 1993. 383 p. Bibliography: p. 374-375. E185.96S225 1993
- Black women in America: An historical encyclopedia. Editor, Darlene Clark Hine. Associate editors, Elsa Barkley Brown, Rosalyn Terborg-Penn. Brooklyn, NY: Carlson Pub., 1993. 2v. (1530 p.) E185.86B542 1993
- See especially entries for aviators, nurses, mathematicians, physicians, and scientists in "Classified list of biographical entries": v.2, p. 1345-1352. Bibliography: p. 1333-1344.
- Chronology of African-American history: significant events and people from 1619 to the present. Hornsby, A. Detroit: Gale Research, 1991. 526 p. Includes bibliographical references. E185.H64. 1991
- Epic lives: one hundred Black women who made a difference. Jessie Carney Smith, editor. Detroit, Visible Ink Press, 1993. 632 p. E185.96.E65 1993 Bibliography: p. 603-632
- Famous firsts of Black Americans. Hancock, S. Gretna, LA: Pelican Pub. Co., 1983. 94 p. Bibliography: p. 93-94. E185.96.H23 1983
- Great women in the struggle: an introduction for young readers. Toyomi Igus, editor. Orange, NJ: Just Us Books, 1991. 107 p. (Book of Black heroes, v. 2) Bibliography: p. 100-101. E185.96.G74 1991
- In Black and white: a guide to magazine articles, newspaper articles, and books concerning more than 15,000 Black individuals and groups. 3rd ed. Spradling, M. M. Detroit: Gale Research Co., c1980. 2 v. (1282 p.) Bibliography: p. 1267-1282. Z1361.N39S655 1980
- Supplement: a guide to magazine articles, newspaper articles, and books concerning more than 6,700 Black individuals and groups. Detroit: Gale Research, c1985. 628 p. Bibliography: p. 621-628. Z1361N39S655 1980 Suppl.
- Notable Black American women. Jessie Carney Smith, editor. Detroit, Gale Research, c1992. 1334 p. Includes bibliographical references. E185.96N68 1992
- **********************************
- **ERIC Clearinghouse for Science, Mathematics, and Environmental Education**

- This digest was funded by the Office of Educational Research and Improvement, U. S. Department of Education under contract no. RR93002013. Opinions expressed in this digest do not necessarily reflect the positions or policies of OERI or the Department of Education.
- This digest is in the public domain and may be freely reproduced. Title: African Americans in Science: Books for Young Readers. ERIC Digest. Document Type: Information Analyses---ERIC Information Analysis Products (IAPs) (071); Information Analyses---ERIC Digests (Selected) in Full Text (073); Available From: ERIC Clearinghouse for Science, Mathematics, and Environmental Education, 1929 Kenny Road, Columbus, OH 43210-1080 (Free). Descriptors: Biographies, Blacks, Books, Educational Resources, Elementary Secondary Education, Multicultural Education, Reading Materials, Science Education Identifiers: African Americans, ERIC Digests

Appendix

Resources

Raymond Webster has a list of more than 30,000 patents and inventions by Black people in USA alone.

* *The Hidden Contributors, Black Scientists and Inventors in America*, by Aaron E. Klein

* *Distinguished African American Scientists of the Twentieth Century*, by James H. Kessler, J.S. Kidd, Renee A. Kidd, Katherine A. Morin

* *20 Years At The Top, A Generation of Black Engineers of the Year*, , by Career Communications Group Publishing

* *Black Scientists of America,* by Richard X. Donovan

* *We Could Not Fail, The First African Americans in the Space Program*, by Richard Paul and Steven Moss

* African-American Inventions That Changed The World: Influential Inventors and Their Revolutionary Creations by Michael A. Carson

Other References

Baker, Henry Edwin. "The Negro as an Inventor." In: Twentieth Century
   Negro Literature, or, A Cyclopedia of Thought on the Vital Topics Relating to the American Negro, edited by D. W. Culp.
   Naperville, IL: J. L. Nichols & Co., 1902. (C, 325.26, C96). Contains a list of several hundred patents issued to African-Americans through the year 1900, a commentary on some of the inventors, a summary of background research, and a brief biography of the author.

Burt, McKinley, Jr. Black Inventors of America. Portland, OR: National
   Book Company, 1989. (C, 926.6048 qB973, 90-13120).

Carnegie, Mary Elizabeth. The Path We Tread; Blacks in Nursing, 1854-1984. Philadelphia: Lippincott, 1986. (C, 331.4816107, C289, 87-053723).

Conner, Douglas L. A Black Physician's Story: Bringing Hope in Mississippi. Jackson, MS: University Press of Mississippi, 1985. (C, 610.924, C752, 88-028184).

Dummett, Clifton Orrin, ed. The Growth and Development of the Negro
   in Dentistry in the United States. Chicago: National Dental Association, 1952. (Z, 617.6, D889).

Haber, Louis. Black Pioneers of Science and Invention. New York: Harcourt, Brace & World, 1970. (Z, 509.2, Ah12).

Hayden, Robert C. and Jacqueline Harris. Nine Black American Doctors. Reading, MA: Addison-Wesley, 1976. (C, 920,H415, 78-24643).

Ho, James K. K., ed. Black Engineers in the United States; A Directory.
   Washington, DC: Howard University Press, 1974. (C, 620.002573, H678, 77-29269).

Ives, Patricia Carter. Creativity and Inventions: The Genius of Afro-Americans and Women in the United States and Their Patents. Arlington, VA: Research Unlimited, 1987. (C, 609.22, I95, 90-14537).

Jenkins, Edward S., ed. American Black Scientists and Inventors. Washington, DC: National Science Teachers Association, 1975. Available as ERIC document. (ED 176-983).

Kenney, John A. The Negro in Medicine. Tuskegee, AL: Tuskegee Institute Press, 1912. (Z, 610.92, Ak).

Manning, Kenneth R. Black Apollo of Science: The Life of Ernest Everett Just. New York: Oxford University Press, 1983. (C, 574.0924, M283, 84-29211).

Organ, Claude H., Jr. and Margaret M. Kosiba. A Century of Black Surgeons; the U.S.A. Experience. Norman, OK: Transcript Press, 1987. (C, 617.0922, c397, 88-033590).

Patterson, Lillie. Benjamin Banneker, Genius of Early America. Nashville: Abingdon, 1978. (C, 520.924, B219, 79-20204).

Pearson, Willie, Jr. and H. Kenneth Bechtel, eds. Blacks, Science and American Education. New Brunswick, NJ: Rutgers University Press, 1989. (C, 508.996073, B631, 89-45053).

Sammons, Vivian O., comp. Blacks in Science and Related Disciplines. LC Science Tracer Bullet TB-85-5. Washington, DC: Library of Congress, Science Reference Section, June 1985. (LC 33.10:85-5).

Stanley, Autumn. "From Africa to America: Black Women Inventors." In: The Technological Woman: Interfacing with Tomorrow, edited by Jan Zimmerman. New York: Praeger, 1983. (C, 305.4, T255, 83-29717).

SPECIAL MATERIALS

Akinsheye, Dexter. African-American Inventors Series. Silver Spring, MD: Three Dimensional Publishing, 1989. A collection of 39 study prints which includes patent diagrams, photographs and histories of some African-American inventors.

U.S. PATENTS

Abrams, William B. Hame Attachment [part for a draft horse's collar],

U.S. Patent 450,550, Apr. 14, 1891.

Alexander, Winser Edward. System for Enhancing Fine Detail in Thermal Photographs, U.S. Patent 3,541,333, Nov. 17, 1970.

Ammons, Virgie M. Fireplace Damper Actuating Tool, U.S. Patent 3,908,633, Sept. 30, 1975.

Ashbourne, Alexander P. Processes for Preparing Cocoanut for Domestic Use, U.S. Patent 163,962, June 1, 1875; Biscuit Cutters, U.S. Patent 170,460, Nov. 30, 1875; Refining Cocoanut Oil, U.S. Patent 230,518, July 27, 1880.

Baker, David. Railway Signal Apparatus, U.S. Patent 1,054,267, Feb. 25, 1913; Signal Apparatus, U.S. Patent 1,154,162, Sept. 21, 1915.

Battle, James. Variable Resistance Resistor Assembly, U.S. Patent 3,691,503, Sept.12, 1972.

Bayless, Robert Gordon. Many patents in field of microencapsulation, including: Solid Globules Containing Dispersed Materials, U.S. Patent 3,922,373, Nov. 25, 1975; Process of Feeding Larval Marine Animals, U.S. Patent 4,073,946, Feb. 14, 1978; Method of Producing Microcapsules and Resulting Product, U.S. Patent 4,107,071, Aug. 15, 1978.

Beard, Andrew Jackson. Rotary Engine, U.S. Patent 478,271, July 5, 1892; Car Coupling ("Jenny Coupler" which reportedly saved the lives of hundreds of trainmen), U.S. Patent 594,059, Nov. 23, 1897.

Belcher, Paul Eugene. Remote AC Power Control with Control Pulses at the Zero Crossing of the AC Wave, U.S. Patent 4,328,482, May 4, 1982.

Bell, Landrow. Smoke Stack for Locomotives, U.S. Patent 115,153, May 23, 1871; Dough Kneader, U.S. Patent 133,823, Dec. 10, 1872.

Benjamin, Lyde W. Broom Moistener and Bridles, U.S. Patent 497,747,
May 16, 1893.

Benjamin, Miriam E. Gong and Signal Chairs for Hotels, etc., U.S. Patent 386,289, July 17, 1888. Adapted for use by the U.S. House of Representatives.

Benton, James W. Lever-derrick, U.S. Patent 658,939, Oct. 2, 1900. According to Henry Baker, Benton walked from Kentucky to Washington carrying his device in order to get his patent.

Bishop, Alfred A. Nuclear Reactor with Self-Orificing Radial Blanket, U.S. Patent 4,077,835, Mar. 7, 1978.

Blackburn, Albert B. Railway Signal, U.S. Patent 376,362, Jan. 10, 1888;
 Spring Seat for Chairs, U.S. Patent 380,420, Apr. 3, 1888; Cash Carrier; U.S. Patent 391,577, Oct. 23, 1888.
Blair, Henry. Designated "colored man" in Patent Office records. Corn Planter, Oct. 14, 1834; Cotton Planter, Aug. 31, 1836. Patents unnumbered during this period.
Blanton, John W. Hydromechanical Rate Damped Servo System, U.S. Patent 3,101,650, Aug. 27, 1963.
Boone, Sarah. Ironing Board, U.S. Patent 473,653, Apr. 26, 1892.
Bowman, Henry A. Making Flags, U.S. Patent 469,395, Feb. 23, 1892.
Boykin, Otis F. Many patents for electronic devices, including Electrical Resistance Element and Method of Making Same, U.S. Patent 3,1911,108, June 22, 1965; Thin Film Capacitor, U.S. Patent 3,394,290, July 23, 1968; Electrical Resistor and Method of Making Same, U.S. Patent 4,561,996, Dec. 31, 1985.
Bradberry, Henrietta. Bed Rack, U.S. Patent 2,320,027, May 25, 1943; Torpedo Discharge Means, U.S. Patent 2,390,688, Dec. 11, 1945.
Brooks, Charles B. Punch, U.S. Patent 507,672, Oct. 31, 1893; Street Sweepers, U.S. Patent 556,711, Mar. 17, 1896; Dust-Proof Bag for Street Sweepers, U.S. Patent 560,154, May 12, 1896.
Brooks, Robert Roosevelt. Line Blanking Apparatus for Color Bar Generating Equipment, U.S. Patent 3,334,178, Aug. 1, 1967; Preset Sensitivity and Amplification Control System, U.S. Patent 3,518,371, June 30, 1970; Vertical and Horizontal Aperture Equalization, U.S. Patent 3,546,372, Dec. 8, 1970.
Brown, Marie Van Brittan. Home Security System Utilizing Television Surveillance, U.S. Patent 3,482,037, Dec. 2, 1969.
Burkins, Eugene. Breech-Loading Cannon, U.S. Patent 649,433, May 15, 1900.
Burnham, Gerald Owens. Direction Coded Digital Stroke Generator Providing A Plurality of Symbols, U.S. Patent 3,938,130, Feb. 10, 1976.
Burr, John A. Lawn Mower, U.S. Patent 624,749, May 9, 1899.
 Burr, William F. Switching Device for Railways, U.S. Patent 636,197, Oct. 31, 1899.
Butler, Francis Edward. Audible Underwater Signal, U.S. Patent 2,803,807, Aug. 20, 1957; Drill Mine, U.S. Patent 2,912,929, Nov. 17, 1959; Watertight Electrical Connector, U.S. Patent 2,991,441, July 4, 1961.
Butler, Richard A. Train Alarm, U.S. Patent 584,540, June 15, 1897.
Byrd, Turner, Jr. Holder for Reins for Harness, U.S. Patent 123,328,

Feb. 6, 1872; Apparatus for Detaching Horses from Carriages, U.S. Patent 124,790, Mar. 19, 1872; Neck Yokes for Wagons, U.S. Patent 126,181, Apr. 30, 1872.

Campbell, William S. Self-Setting Animal Trap, U.S. Patent 246,369, Aug. 30, 1881.

Carruthers, George R. Image Converter for Detecting Electro-Magnetic Radiation Especially in Short Wave Lengths, U.S. Patent 3,478,216, Nov. 11, 1969. Dr. Carruthers designed the ultraviolet camera/spectrograph which was adapted for use on the lunar surface and placed there by the Apollo 16 astronauts.

Carter, William C. Umbrella Stand, U.S. Patent 323,397, Aug. 4, 1885.

Carver, George Washington. Carver refused to patent a vast majority of the hundreds of products he developed while at Tuskegee Institute. Patent Office records show the following: Cosmetic and Producing the Same, U.S. Patent 1,522,176, Jan. 6, 1925; Paint and Stain and Producing the Same, U.S. Patent 1,541,478, June 9, 1925; Producing Paints and Stains, U.S. Patent 1,632,365, June 14, 1927.

Certain, Jerry M. Parcel Carrier for Bicycles, U.S. Patent 639,708, Dec. 26, 1899.

Chappelle, Emmett W. Numerous patents in photobiology, including Light Detection Instrument, U.S. Patent 3,520,660, July 14, 1970; Method of Detecting and Counting Bacteria, U.S. Patent 3,971,703, July 27, 1976; Rapid, Quantitative Determination of Bacteria in Water, U.S. Patent 4,385,113, May 24, 1983.

Cherry, Matthew A. Velocipede, U.S. Patent 382,351, May 8, 1888; Street Car Fender, U.S. Patent 531,908, Jan. 1, 1895.

Christian, John B. Many patents in chemical engineering, including Grease Composition for Use at High Temperatures and High Speeds, U.S. Patent 3,518,189, June 30, 1970; Fluorine-Containing Benzimidazoles, U.S. Patent 4,267,348, May 12, 1981; Perfluoroalkylether Substituted Phenyl Phosphines, U.S. Patent 4,454,349, June 12, 1984.

Church, Titus S. Carpet Beating Machine, U.S. Patent 302,237, July 22, 1884.

Cook, George. Automatic Fishing Device, U.S. Patent 625,829, May 30, 1899.

Cooper, James. Elevator Device, U.S. Patent 536,605, Apr. 2, 1895; Elevator Safety Device, U.S. Patent 590,257, Sept. 21, 1897.

Cooper, John Richard. Chemical patents include Two-Stage Phosgenation Process for Preparing Aromatic Isocyanates, U.S.

Patent 3,234,253, Feb. 8, 1966; Process for Isolating a Fluorine-Containing Polymer, U.S. Patent 3,536,683, Oct. 27, 1970; Separation of Distillable Isocyanates from their Phosgenation Masses, U.S. Patent 3,694,323, Sept. 26, 1972.

Cooper, Jonas. Shutter and Fastening Therefor, U.S. Patent 276,563, May 1, 1883.

Cotton, Donald J. Vertical Liquid Electrode Employed in Electrolytic Cells, U.S. Patent 4,040,932, Aug. 9, 1977; Capillary Liquid Fuel Nuclear Reactor, U.S. Patent 4,327,443, Apr. 27, 1982.

Cowans, Beatrice L. and Virginia E. Hall. Embroidered Fruit Bowl Wall Hanging and Kit, U.S. Patent 4,016,314, Apr. 5, 1977.

Cralle, Alfred L. Ice-Cream Mold and Disher, U.S. Patent 576,395, Feb. 2, 1897.

Creamer, Henry. Seven patents for steam trap devices, including Steam Feed Water Trap, U.S. Patent 313,854, Mar. 17, 1885; Steam Feed Water Traps, U.S. Patent 358,964, Mar. 8, 1887; Steam Trap and Feeder, U.S. Patent 394,463, Dec. 11, 1888.

Crossley, Frank Alphonso. Patents in metallurgical engineering include Titanium Base Alloy, U.S. Patent 2,798,807, July 9, 1957; Grain Refinement of Beryllium with Tungsten Carbide and Titanium Diboride, U.S. Patent 3,117,001, Jan. 7, 1964; Grain Refinement of Titanium Alloys, U.S. Patent 4,420,460, Dec. 13, 1983.

Crosthwait, David Nelson Jr. Thirty-nine patents, authority on heat transfer, ventilation and air conditioning; helped design Radio City Music Hall heating system. Titles include Apparatus for Returning Water to Boilers, U.S. Patent 1,353,457, Sept. 21, 1920; Method and Apparatus for Setting Thermostats, U.S. Patent 1,661,323, Mar. 6, 1928; Differential Vacuum Pump, U.S. Patent 1,755,430, Apr. 22, 1930.

Dacons, Joseph Carl. Numerous patents in organic chemistry, including Process for the Manufacture of Nitroform and Its Salts, U.S. Patent 3,125,606, Mar. 17, 1964; Dodecanitroquaterphenyl, U.S. Patent 3,450,778, June 17, 1969; Recrystallization of Hexanitrostilbene from Nitric Acid and Water, U.S. Patent 4,260,847, Apr. 7, 1981.

Davidson, Shelby J. Paper-Rewind Mechanism for Adding Machines, U.S. Patent 884,721, Apr. 14, 1908.

Davis, William R., Jr. Library Table, U.S. Patent 208,378, Sept. 24, 1878; Game Table, U.S. Patent 362,611, May 10, 1887.

Deitz, William A. Shoe [method of constructing boots], U.S. Patent 64,205, Apr. 30, 1867.

Dent, Anthony L. Rehydrated Silica Gel Dentifrice Abrasive, U.S. Patent 4,346,071, Aug. 24, 1982; Toothpaste Containing pH-Adjusted Zeolite, U.S. Patent 4,349,533, Sept. 14, 1982.

Dickinson, Joseph Hunter. Many patents for player piano machinery, including Adjustable Tracker for Pneumatic Playing Attachments, U.S. Patent 915,942, Mar. 23, 1909; Volume-Controlling Means for Mechanical Musical Instruments, U.S. Patent 926,178, June 29, 1909; Player-Piano, U.S. Patent 1,028,996, June 11, 1912.

Diuguid, Lincoln Isaiah. Burning Efficiency Enhancement Method; Alkynol Fuel Additive for Improved Automobile Mileage, U.S. Patent 4,539,015, Sept. 3, 1985.

Dorman, Linneaus Cuthbert. Prolific inventor with many chemical patents, including 3,5-Dihalo-4-Cyanoalkoxy Phenols, U.S. Patent 3,468,926, Sept. 23, 1969; Absorbents for Airborne Formaldehyde, U.S. Patent 4,517,111, May 14, 1985; Composites of Unsintered Calcium Phosphates and Synthetic Biodegradable Polymers Useful as Hard Tissue Prosthetics, U.S. Patent 4,842,604, June 27, 1989.

Dorticus, Clatonia J. Device for Applying Coloring Liquids to Sides of Soles or Heels of Shoes, U.S. Patent 535,820, Mar. 19, 1895; Machine for Embossing Photographs, U.S. Patent 537,442, Apr. 16, 1895; Photographic Print Wash, U.S. Patent 537,968, Apr. 23, 1895.

Douglass, William. Several patents in harvesting equipment, including Self-Binding Harvester, U.S. Patent 789,010, May 2, 1905; Band-Twister, U.S. Patent 789,120, May 2, 1905; Carrier Chain, U.S. Patent 789,122, May 2, 1905.

Doyle, James. Automatic Serving System, U.S. Patent 1,019,137, Mar. 5,
1912.

Dugger, Cortland Otis. Inventor of "Duggerite" (barium magnesium aluminate). Patents include Method for Growing Single Oxide Crystals, U.S. Patent 3,595,803, July 27, 1971; Solid-State Laser Produced by a Chemical Reaction Between a Germanate and an Oxide Dopant, U.S. Patent 3,624,547, Nov. 30, 1971; Aluminum Nitride Single Crystal Growth from a Molten Mixture with Calcium Nitride, U.S. Patent 3,933,573, Jan. 20, 1976.

Dunnington, James H. Horse Detacher, U.S. Patent 578,979, Mar. 16, 1897.

Elder, Clarence L. Timing Device, U.S. Patent 3,165,188, Jan. 12, 1965;

Non-Capsizable Container, U.S. Patent 3,367,525, Feb. 6, 1968; Bidirectional Monitoring and Control System, U.S. Patent 4,000,400, Dec. 28, 1976.
Elkins, Thomas. Dining and Ironing Table and Quilting Frame, U.S. Patent 100,020, Feb. 22, 1870; Chamber Commode, U.S. Patent 122,518, Jan. 9, 1872; Refrigerating Apparatus, U.S. Patent 221,222, Nov. 4, 1879.
Ferrell, Frank J. Eight patents for improvements in valves for steam engines, beginning with Valve, U.S. Patent 428,621, May 27, 1890.
Fisher, David A., Jr. Joiner's Clamp, U.S. Patent 162,281, Apr. 20, 1875; Furniture Caster, U.S. Patent 174,794, Mar. 14, 1876.
Flemmings, Robert F., Jr. Guitar, U.S. Patent 338,727, Mar. 3, 1886.
Frye, Clara. Surgical Appliance, U.S. Patent 847,758, Mar. 19, 1907.
Gant, Virgil Arnett. Method of Treating Hair, U.S. Patent 2,643,375, June 23, 1953; Hair Treating Composition and Method of Use for Setting, U.S. Patent 2,750,947, Sept. 4, 1956; Ammonium Polysiloxanolate Hair Treating Composition and Method for Using Same, U.S. Patent 2,787,272, Apr. 2, 1957.
Gaskin, Frances C. Sun Protectant Composition and Method, U.S. Patent 4,806,344, Feb. 21, 1989.
Gay, Eddie Charles. Cathode for a Secondary Electrochemical Cell, U.S. Patent 3,907,589, Sept. 23, 1975; Method of Preparing Electrodes with Porous Current Collector Structures and Solid Reactants for Secondary Electrochemical Cells, U.S. Patent 3,933,520, Jan. 20, 1976; Compartmented Electrode Structure, U.S. Patent 4,029,860, June 14, 1977.
Goode, Sarah E. Folding Cabinet Bed, U.S. Patent 322,177, July 14, 1885.
Gourdine, Meredith C. Over 23 U.S. patents for electrogasdynamic devices. These include such titles as Electrogasdynamic Systems and Methods, U.S. Patent 3,582,694, June 1, 1971; Electrogasdynamic Coating System, U.S. Patent 4,574,092, Mar. 4, 1986; Method for Airport Fog Precipitation, U.S. Patent 4,671,805, June 9, 1987.
Grant, George F. Golf-Tee, U.S. Patent 638,920, Dec. 12, 1899.
Green, Harry James, Jr. Method of Making a Striated Support for Filaments, U.S. Patent 3,548,045, Dec. 15, 1970; Substrate for Mounting Filaments in Close-Spaced Parallel Array, U.S. Patent 3,584,130, June 8, 1971; Method for Sealing Microelectronic Device Packages, U.S. Patent 3,648,357, Mar. 14, 1972.
Greene, Frank S., Jr. Use of Faulty Storage Circuits by Position

Coding, U.S. Patent 3,654,610, Apr. 4, 1972.
Grenon, Henry. Razor Stropping Device, U.S. Patent 554,867, Feb. 18, 1896.
Griffin, Bessie V. Portable Receptacle Support, U.S. Patent 2,550,554, Apr. 24, 1951.
Hall, Lloyd Augustus. More than one hundred patents worldwide in food chemistry. Discovered curing salts for meat-packing. Patents include Synergistic Antioxidants and the Methods of Preparing Same, U.S. Patent 2,493,288, Jan. 3, 1950; Meat-curing Salt Composition, U.S. Patent 2,770,551, Nov. 13, 1956; Method of Preserving Fresh Frozen Pork Trimmings, U.S. Patent 2,845,358, July 29, 1958.
Hammond, Julia T. Apparatus for Holding Yarn Skeins, U.S. Patent 572,985, Dec. 15, 1896.
Harrison, Emmett Scott. Gas Turbine Air Compressor and Control Therefor, U.S. Patent 3,606,971, Sept. 21, 1971; Turbojet Afterburner Engine with Two-Position Exhaust Nozzle, U.S. Patent 4,242,865, Jan. 6, 1981.
Hawkins, Walter Lincoln. Numerous chemical patents, including Preparation of 1,2,Di-Primary Amines, U.S. Patent 2,587,043, Feb. 26, 1952; Stabilized Straight-Chain Hydrocarbons, U.S. Patent 2,889,306, June 2, 1959; Stabilized Alpha-Mono-Olefinic Polymers, U.S. Patent 3,304,283, Feb. 14, 1967.
Hearns, Robert. Sealing Attachment for Bottles, U.S. Patent 598,929, Feb. 15, 1898; Detachable Car Fender, U.S. Patent 628,003, July 4, 1899. [In Baker list as "Hearness"].
Henderson, Henry Fairfax, Jr. Weight Loss Control System, U.S. Patent
4,111,336, Sept. 5, 1978.
Hill, Henry Aaron. Patents include Manufacture of Azodicarbonamide, U.S. Patent 2,988,545, June 13, 1961; Foamable Composition Comprising a Thermoplastic Polymer and Barium Azocarbonate and Method of Foaming, U.S. Patent 3,141,002, July 14, 1964; Curing Furfuryl-Alcohol-Modified Urea Formaldehyde Condensates, U.S. Patent 3,297,611, Jan. 10, 1967.
Hilyer, Andrew F. Water Evaporator Attachment for Hot Air Registers, U.S. Patent 435,095, Aug. 26, 1890; Evaporator for Hot Air Registers, U.S. Patent 438,159, Oct. 14, 1890.
Hodge, John E. Numerous patents in the area of food technology, including Novel Reductones and Methods of Making Them, U.S. Patent 2,936,308, May 10, 1960; Glucose-Amine Sequestrants, U.S. Patent 2,996,449, Aug. 15, 1961; Substituted Benzodioxan

Sweetening Compound, U.S. Patent 4,146,650, Mar. 27, 1979.
Holmes, Lydia M. Knockdown Wheeled Toy, U.S. Patent 2,529,692, Nov. 14, 1950.
Howard, Darnley Moseley. Method of Making Radome with an Integral
Antenna, U.S. Patent 3,451,127, June 24, 1969.
Hunter, John W. Portable Weighing Scale, U.S. Patent 570,553, Nov. 3,
1896.
Hyde, Robert N. Composition for Cleaning and Preserving Carpets, U.S. Patent 392,205, Nov. 6, 1888.
Jackson, Benjamin F. Heating Apparatus, U.S. Patent 599,985, Mar. 1, 1898; Matrix Drying Apparatus, U.S. Patent 603,879, May 10, 1898; Gas Burner, U.S. Patent 622,482, Apr. 4, 1899.
Jackson, William H. Railway Switch, U.S. Patent 578,641, Mar. 9, 1897;
Automatic Locking Switch, U.S. Patent 609,436, Aug. 23, 1898.
Johnson, Andrew R. Precision Digital Delay Circuit, U.S. Patent 3,376,436, Apr. 2, 1968.
Johnson, Daniel. Rotary Dining Table, U.S. Patent 396,089, Jan. 15, 1889; Lawn Mower Attachment, U.S. Patent 410,836, Sept. 10, 1889; Grass Receivers for Lawn Mowers, U.S. Patent 429,629, June 10, 1890.
Johnson, John ("Jack") Arthur. A little-known footnote to the career of the world's first black heavyweight champion, subject of Sackler's play, The Great White Hope; two patents applied for while he was a prisoner at Leavenworth: Wrench, U.S. Patent 1,413,121, Apr. 18, 1922, and Theft-Preventing Device for Vehicles, U.S. Patent 1,438,709, Dec. 12, 1922.
Johnson, Powell. Eye Protector, U.S. Patent 234,039, Nov. 2, 1880.
Johnson, Willie H. Mechanism for Overcoming Dead Centers, U.S. Patent 554,223, Feb. 4, 1896 and U.S. Patent 612,345, Oct. 11, 1898.
Johnson, Willis. Eggbeater, U.S. Patent 292,821, Feb. 5, 1884.
Jones, Frederick McKinley. Over 60 patents, including
Ticket-dispensing Machine, U.S. Patent 2,163,754, June 27, 1939; Air Conditioning Unit, U.S. Patent 2,475,841, July 12, 1949; Starter Generator, U.S. Patent 2,475,842, July 12, 1949. McKinley's inventions made it possible to refrigerate trucks for long-distance travel.
Jones, Howard St. Claire, Jr. Many patents in the area of microwave and other types of antennas, including Electronically Scanned

Microwave Antennas, U.S. Patent 3,268,901, Aug. 23, 1966; Dielectric-Loaded Antenna with Matching Window, U.S. Patent 3,518,683, June 30, 1970; Multifrequency Series-Fed Edge Slot Antenna, U.S. Patent 4,305,078, Dec. 8, 1981.

Jones, John Leslie. Numerous patents, including Preparation of Substituted Phenols, U.S. Patent 2,497,503, Feb. 14, 1950; Personnel Restraint System for Vehicular Occupants, U.S. Patent 3,690,695, Sept. 12, 1972; Smokeless Slow Burning Cast Propellant, U.S. Patent 4,112,849, Sept. 12, 1978.

Joyce, James A. Coal or Ore Bucket, U.S. Patent 603,143, Apr. 26, 1898.

Julian, Percy Lavon. Prolific researcher in organic chemistry and synthesizer of the drugs physostigmine and cortisone; holder of over 100 patents worldwide. Named to National Inventors Hall of Fame in 1990. Titles include Preparation of Cortisone, U.S. Patent 2,752,339, June 26, 1956; 16-Aminomethyl-17-Alkyltestosterone Derivatives, U.S. Patent 3,149,132, Sept. 15, 1964; Method for Preparing 16(Alpha)-Hydroxypregnenes and Intermediates Obtained Therein, U.S. Patent 3,274,178, Sept. 20, 1966.

Kenner, Mary Beatrice. Carrier Attachment for Invalid Walkers, U.S. Patent 3,957,071, May 18, 1976; Bathroom Tissue Holder, U.S. Patent 4,354,643, Oct. 19, 1982; Shower Wall and Bathtub Mounted Back Washer, U.S. Patent 4,696,068, Sept. 29, 1987.

Knox, Lawrence Howland. Production of Arecoline, U.S. Patent 2,506,458, May 2, 1950; Photochemical Preparation of Tropilidenes, U.S. Patent 2,647,081, July 28, 1953.

Knox, William J., Jr. At least 25 patents related to photography between 1950 and 1970, including Coating Aids for Gelatin Compositions, U.S. Patent 3,038,804, June 2, 1962; Gelatin Coating Compositions, U.S. Patent 3,306,749, Feb. 28, 1967; Coating Aids for Hydrophilic Colloid Layers of Photographic Elements, U.S. Patent 3,539,352, Nov. 10, 1970.

Latimer, Lewis Howard. One of the "Edison Pioneers" with patents in electricity, Latimer was creative in other fields as well. Patents include Manufacturing Carbons, U.S. Patent 252,386, Jan. 17, 1882; Apparatus for Cooling and Disinfecting, U.S. Patent 334,078, Jan. 12, 1886; Locking Racks for Hats, Coats, Umbrellas, etc., U.S. Patent 557,076, Mar. 24, 1896. See "Books" section for Latimer's publication on the Edison system.

Lavalette, William A. Printing Press, U.S. Patent 208,184, Sept. 17, 1878.

Lee, Joseph. Kneading Machine, U.S. Patent 524,042, Aug. 7, 1894; Bread Crumbing Machine, U.S. Patent 540,553, June 4, 1895.

LeVert, Francis Edward. Numerous patents in physics, including Threshold Self-Powered Gamma Detector for Use as a Monitor of Power in a Nuclear Reactor, U.S. Patent 4,091,288, May 23, 1978; Monitor for Deposition on Heat Transfer Surfaces, U.S. Patent 4,722,610, Feb. 2, 1988; Continuous Level Fluid Detector, U.S. Patent 4,805,454, Feb. 21, 1989.

Lewis, Edward R. Spring Gun, U.S. Patent 362,096, May 3, 1887.

Lewis, James Earl. Antenna Feed for Two Coordinate Tracking Radars, U.S. Patent 3,388,399, June 11, 1968.

Linden, Henry. Piano Truck, U.S. Patent 459,365, Sept. 8, 1891.

Loudin, Frederick J. Sash Fastener, U.S. Patent 510,432, Dec. 12, 1893; Key Fastener, U.S. Patent 512,308, Jan. 9, 1894.

Love, John L. Plasterer's Hawk, U.S. Patent 542,419, July 9, 1895; Pencil Sharpener, U.S. Patent 594,114, Nov. 23, 1897.

Lu Valle, James E. Photographic Processes, U.S. Patent 3,219,445, Nov. 23, 1965; Photographic Medium and Methods of Preparing Same, U.S. Patent 3,219,448, Nov. 23, 1965; Sensitizing Photographic Media, U.S. Patent 3,219,451, Nov. 23, 1965.

Madison, Shannon L. Refrigerating Apparatus, U.S. Patent 3,208,232, Sept. 28, 1965; Electrical Wiring Harness Termination System, U.S. Patent 4,793,820, Dec. 27, 1988.

Maloney, Kenneth Morgan. Alumina Coatings for an Electric Lamp, U.S. Patent 3,868,266, Feb. 25, 1975; Alumina Coatings for Mercury Vapor Lamps, U.S. Patent 4,079,288, Mar. 14, 1978.

Martin, Thomas J. Fire Extinguisher, U.S. Patent 125,063, Mar. 26, 1872. [In Baker list as "Marshall."]

Matzeliger, Jan Earnst. Credited with revolutionizing the American shoe industry with his shoe-lasting machine. Patents include Nailing Machine, U.S. Patent 421,954, Feb. 25, 1890; Tack Separating and Distributing Mechanism, U.S. Patent 423,937, Mar. 25, 1890; Lasting Machine, U.S. Patent 459,899, Sept. 22, 1891.

McCoy, Elijah. Twenty-eight patents for various inventions through 1900, and many thereafter, most in the field of automatic lubrication. The term for the genuine article, "The Real McCoy", is associated with his work. Titles include Lubricator for Steam Engines, U.S. Patent 129,843, July 23, 1872; Steam Cylinder Lubricator, U.S. Patent 173,032, Feb. 1, 1876; Lawn Sprinkler

Design, U.S. Patent 631,549, Sept. 26, 1899.

McCree, Daniel. Portable Fire Escape, U.S. Patent 440,322, Nov. 11, 1890.

Miles, Alexander. Elevator, U.S. Patent 371,207, Oct. 11, 1887.

Millington, James E. Thermostable Dielectric Material, U.S. Patent 3,316,178, Apr. 25, 1967; Method of Making Expandable Styrene-Type Beads, U.S. Patent 4,286,069, Aug. 25, 1981; Method of Making Styrene-Type Polymer, U.S. Patent 4,730,027, Mar. 8, 1988.

Morgan, Garrett Augustus. Breathing Device, U.S. Patent 1,113,675, Mar. 24, 1914; Three-Way Traffic Signal, U.S. Patent 1,475,024, Nov. 20, 1923. Morgan used his gas mask in 1916 during the heroic rescue of workers trapped in a tunnel disaster.

Morris, Joel Morton. Switching System Charging Arrangement, U.S. Patent 3,688,047, Aug. 29, 1972.

Murray, George W. Congressman from South Carolina (see entry for Congressional Record in "Selected Periodical Articles"). Eight patents for agricultural implements in 1894, including Combined Furrow Opener and Stalk Knocker, U.S. Patent 517,960, Apr. 10, 1894; Reaper, U.S. Patent 520,892, June 5, 1894; Cotton Chopper, U.S. Patent 520,888, June 5, 1894.

Nash, Henry H. Life Preserving Stool, U.S. Patent 168,519, Oct. 5, 1875.

Neblett, Richard Flemon. Several chemical patents, including Gasoline Composition, U.S. Patent 2,955,928, Oct. 11, 1960; Motor Fuel Composition, U.S. Patent 3,054,666, Sept. 18, 1962; Oil-Soluble Ashless Dispersant-Detergent-Inhibitors, U.S. Patent 3,511,780, May 12, 1970.

Newman, Lyda D. Brush, U.S. Patent 614,335, Nov. 15, 1898.

Nickerson, William J. Piano Attachment, U.S. Patent 627,739, June 27, 1899.

Outlaw, John W. Horseshoe, U.S. Patent 614,273, Nov. 15, 1898.

Parker, Alice H. Heating Furnace, U.S. Patent 1,325,905, Dec. 23, 1919.

Parsons, James A., Jr. Numerous patents in metallurgy, including Iron Alloy, U.S. Patent 1,728,360, Sept. 17, 1929; Making Silicon Iron Compounds, U.S. Patent 1,819,479, Aug. 18, 1931; Corrosion-Resisting Ferrous Alloy, U.S. Patent 2,200,208, May 7, 1940.

Perryman, Frank R. Caterer's Tray Table, U.S. Patent 468,038, Feb. 2, 1892.

Peterson, Henry. Attachment for Lawn Mowers, U.S. Patent 402,189,

Apr. 30, 1889.

Phelps, William H. Apparatus for Washing Vehicles, U.S. Patent 579,242, Mar. 23, 1897.

Pickering, John F. Air Ship, U.S. Patent 643,975, Feb. 20, 1900.

Pope, Jessie T. Croquignole Iron, U.S. Patent 2,409,791, Oct. 22, 1946. Patented with the aid of Eleanor Roosevelt, according to Autumn Stanley (see under "Books").

Porter, James Hall. Gas Well Sulfur Removal by Diffusion Through Polymeric Membranes, U.S. Patent 3,534,528, Oct. 20, 1970.

Prince, Frank Rodger. Production of 2-Pyrrolidones, U.S. Patent 3,637,743, Jan. 25, 1972.

Pugsley, Samuel. Gate Latch, U.S. Patent 357,787, Feb. 15, 1887.

Purvis, William B. Sixteen patents 1882-1897, ten for paper bag machines, as in U.S. Patent 293,353, Feb. 12, 1884; also Magnetic Car Balancing Device, U.S. Patent 539,542, May 21, 1895; Electric Railway System, U.S. Patent 588,176, Aug. 17, 1897.

Ransom, Victor Llewellyn. Traffic Data Processing System, U.S. Patent
3,231,866, Jan. 25, 1966; Method and Apparatus for Gathering Peak Load Traffic Data, U.S. Patent 3,866,185, Feb. 11, 1975.

Reynolds, Humphrey H. Window Ventilator for Railroad Cars, U.S. Patent 275,271, Apr. 3, 1883; Safety Gate for Bridges, U.S. Patent 437,937, Oct. 7, 1890.

Reynolds, Robert R. Non-Refillable Bottle, U.S. Patent 624,092, May 2,
1899.

Rhodes, Jerome B. Water Closet, U.S. Patent 639,290, Dec. 19, 1899.

Richardson, Albert C. Several patents, including Hame Fastener, U.S. Patent 255,022, Mar. 14, 1882; Casket Lowering Device, U.S. Patent 529,311, Nov. 13, 1894; Insect Destroyer, U.S. Patent 620,362, Feb. 28, 1899.

Richardson, William H. Cotton Chopper, U.S. Patent 343,140, June 1, 1886; Child's Carriage, U.S. Patent 405,599, June 18, 1889; also U.S. Patent 405,600, same title and date.

Richey, Charles V. Several patents in different areas of invention, including Fire Escape Bracket, U.S. Patent 596,427, Dec. 28, 1897; Combined Hammock and Stretcher, U.S. Patent 615,907, Dec. 13, 1898; Telephone Register and Lock-Out Device, U.S. Patent 1,063,599, June 3, 1913.

Rillieux, Norbert. Rillieux's inventions completely changed the sugar industry. Sugar Works, U.S. Patent 3,237, Aug. 26, 1843; Sugar

Making, Improvement in, U.S. Patent 4,879, Dec. 10, 1846.

Roberts, Louis W. Gaseous Discharge Device, U.S. Patent 3,072,865, Jan. 8, 1963; GASAR (Device for Gas Amplification by Stimulated Emission and Radiation), U.S. Patent 3,257,620, June 21, 1966; Gallium-Wetted Movable Electrode Switch, U.S. Patent 3,377,576, Apr. 9, 1968.

Robinson, Elbert R. Electric Railway Trolley, U.S. Patent 505,370, Sept. 19, 1893; Casting Composite or Other Wheels, U.S. Patent 594,286, Nov. 23, 1897.

Rose, Raymond E. Control Apparatus [Aerodynamic], U.S. Patent 3,618,388, Nov. 9, 1971.

Russell, Edwin Roberts. Eleven patents on atomic energy processes, including The Separation of Plutonium from Uranium and Fission Products, U.S. Patent 2,855,269, Oct. 7, 1958; Ion Exchange Adsorption Process for Plutonium Separation, U.S. Patent 2,992,249, July 11, 1961; Thorium-Oxide or Thorium-Uranium Oxide with Magnesium Oxide, U.S. Patent 3,309,323, Mar. 14, 1967.

Ryder, Earl. High Silicon Cast Iron, U.S. Patent 3,129,095, Apr. 14, 1964.

Sampson, George T. Sled Propeller, U.S. Patent 312,388, Feb. 17, 1885;
Clothes Drier, U.S. Patent 476,416, June 7, 1892.

Sampson, Henry Thomas. Binder System for Propellants and Explosives, U.S. Patent 3,140,210, July 7, 1964; Case Bonding System for Cast Composite Propellants, U.S. Patent 3,212,265, Oct. 19, 1965; High Voltage Gamma Electric Cell, U.S. Patent 3,591,860, July 6, 1971.

Sams, Adolphus. Multiple Stage Rocket, U.S. Patent 3,199,455, Aug. 10,
1965; Emergency Release for Extraction Chute, U.S. Patent 3,257,089, June 21, 1966; Rocket Motor Fuel Feed, U.S. Patent 3,310,938, Mar. 28, 1967.

Scottron, Samuel R. Several patents, including Adjustable Window Cornice, U.S. Patent 224,732, Feb. 17, 1880; Pole Tip, U.S. Patent 349,525, Sept. 21, 1886; Curtain Rod, U.S. Patent 481,720, Aug. 30, 1892. See Scottron's article on "Manufacturing Household Articles" in "Selected Periodical Articles" section of this bibliography.

Shaw, Earl D. Free-Electron Amplifier Device with Electromagnetic Radiation Delay Element, U.S. Patent 4,5299,942, July 16, 1985.

Silvera, Esteban. Ram-Valve Level Indicator, U.S. Patent 3,718,157,

Feb. 27, 1973.

Smartt, Brinay. Reversing-Valve, U.S. Patent 799,498, Sept. 12, 1905; Valve Gear, U.S. Patent 935,169, Sept. 28, 1909; Wheel, U.S. Patent 1,052,290, Feb. 4, 1913.

Smith, Mildred E. Family Relationships Card Game, U.S. Patent 4,230,321, Oct. 28, 1980.

Smoot, Lanny S. Numerous patents in communications technology, including Optical Receiver Circuit with Active Equalizer, U.S. Patent 4,565,974, Jan. 21, 1986; Teleconference Facility with High Resolution Video Display [Bellcore's "Videowindow"], U.S. patent 4,890,314, Dec. 26, 1989; Teleconferencing Terminal with Camera Behind Display Screen, U.S. Patent 4,928,301, May 22, 1990.

Spikes, Richard B. Automatic Gear Shift, U.S. Patent 1,889,814, Dec. 6, 1932; Transmission and Shifting Means Therefor, U.S. Patent 1,936,996, Nov. 28, 1933; Automatic Safety Brake System, U.S. Patent 3,015,522, Jan 2, 1962.

Stanard, John. Oil Stove, U.S. Patent 413,689, Oct. 29, 1889; Refrigerator, U.S. Patent 455,891, July 14, 1891. [In Baker's list as "Standard."]

Stewart, Albert Clifton. Redox Couple Radiation Cell, U.S. Patent 3,255,044, June 7, 1966; Electric Cell, U.S. Patent 3,255,045, June 7, 1966.

Stewart, Marvin Charles. Arithmetic Unit for Digital Computers, U.S. Patent 3,395,271, July 30, 1968; System for Interconnecting Electrical Components, U.S. Patent 3,605,063, Sept. 14, 1971.

Stokes, Rufus. Exhaust Purifier, U.S. Patent 3,378,241, Apr. 16, 1968; Air Pollution Control Device, U.S. Patent 3,520,113, July 14, 1970.

Taylor, Moddie Daniel. Preparation of Anhydrous Alkaline Earth Halides, U.S. Patent 2,801,899, Aug. 6, 1957; Ion Exchange Adsorption Process for Plutonium Separation, U.S. Patent 2,992,249, July 11, 1961; Preparation of Anhydrous Lithium Salts, U.S. Patent 3,049,406, Aug. 14, 1962.

Turner, Madeline M. Fruit Press, U.S. Patent 1,180,959, Apr. 25, 1916.

Walker, M. Lucius, Jr. Laminar Fluid NOR Element, U.S. Patent 3,478,764, Nov. 18, 1969.

Weaver, Rufus J. Stairclimbing Wheelchair, U.S. Patent 3,411,598, Nov. 19, 1968.

West, James Edward. Over 100 patents worldwide in acoustical physics.

Titles include Techniques for Fabrication of Foil Electret, U.S. Patent 3,945,112, Mar. 23, 1976; Technique for Removing Surface and Volume Charges from Thin High Polymer Films, U.S. Patent 4,248,808, Feb. 3, 1981; Noise Reducing Processing Arrangement for Microphone Arrays, U.S. Patent 4,802,227, Jan. 31, 1989.

Woods, Granville T. Known as the "Black Edison," with more than 60 patents, including Telephone Transmitter, U.S. Patent 308,817, Dec. 2, 1884; Electro-Motive Railway System, U.S. Patent 385,034, June 26, 1888; Amusement Apparatus, U.S. Patent 639,692, Dec. 19, 1899.

There's No Place Like Home

BY: Kate Nkansa   DATE: 1 October, 2010

"Africa will overcome the challenges of poverty, underdevelopment and global marginalisation not because of its wealth in natural resources, but because of its intellectual ability properly to manage and utilise these resources for the benefit of the peoples of our continent. In this sense the resources embedded in earth Africa may turn into a curse if Africa does not develop the intellectual capital to empower the African masses and the governments they elect to exploit these resources for the greater good of the citizen." Taken from Thabo Mbeki's speech at the All African Students union Conference.

Mbeki's remarks should include young professional Africans who are part of the workforce in developed countries. I am talking about the 25-45 year age group of Africans in the Diaspora. I have been doing some research concerning just how many young African professionals are working in first world countries. The statistics are astonishing to say the least. According to the Organisation for Economic Co-operation and Development (OECD), which includes most of the world's richest nations, 25% -50% of college-educated citizens of African countries including Ghana, Mozambique, Kenya and Uganda lived in an OECD country. In contrast, less than 5% of the skilled citizens of newly emerging powers such as India, China and Brazil live abroad.

Africa's problems cannot fully rest on the shoulders of our leadership. We all need to be accountable for the state of our continent. Based on the information, our intellectual capital is virtually non-existent in Africa.

Since 2000, it is estimated that Africa loses 20 000 professionals annually. 40 000 doctoral graduates reside outside the continent. Statistics show that Africa's technical and managerial talent living outside the continent has more than doubled in a generation. The IOM World Migration Report 2005 estimated that there are over 21 000 Nigerian doctors practicing in the USA and over 200 000 African scientists in the United States, more than on the entire African continent.

In 2003, 5,880 UK work permits were approved for health and medical personnel from South Africa, 2,825 from Zimbabwe, 1,510 from Nigeria and 850 from Ghana. According to a report by the British Medical Journal, between 1993 and 2002 Ghana lost 630 medical doctors, 410 pharmacists, 87 laboratory technicians and 11 325 nurses.

Help Minorities Succeed in STEM Education

October 19, 2010 by *Kristin Drouin*

Expanding Underrepresented Minority Participation: America's Science and Technology Talent at the Crossroads

National Academies

Although minorities are the fastest-growing subset of the U.S. population, they remain considerably underrepresented in science and technology professions, according to this report. It says that African-Americans, Native Americans and Hispanics account for 28.5 percent of the population but only 9.1 percent of college-educated workers employed in the science or engineering sectors.

With the minority population increasing and the size of the science and engineering workforce expected to expand faster than any other sector in the next few years, the report said that attracting more workers from minority groups is crucial to maintaining forward momentum and development in these industries.

The authors say that while the United States is a world leader in science, engineering and technological development, the country stands at a "crossroads" that could be addressed by putting more emphasis on education. A UCLA study cited in the report found that the number of underrepresented minority college students who plan to major in STEM [science, technology, engineering or mathematics] fields is comparable to their Asian-American and white peers – yet their degree completion rate is considerably lower, with percentages of 24-year-old minority students who have obtained a STEM undergraduate degree ranging from 2.2 to 3.3 percent.

Describing these students as "a vastly underused resource and a lost opportunity for meeting our nation's technology needs," the authors recommend that the federal government, industry representatives and postsecondary institutions partner with K-12 schools to promote interest in, access to and the affordability of postsecondary STEM curriculum and training.

Business Week
February 15, 2008

Bridging Engineering's Minority Gap

Encouraging more women, African Americans, Latinos, and people with disabilities to pursue math and science makes good business sense

by John E. Kelly III

As Presidents Day marks the start of National Engineers Week, it's worth wondering what Presidents George Washington and Abraham Lincoln might have said about the state of engineering today.

As it happens, both Washington and Lincoln were land surveyors—the prototypical engineer of their day—and Lincoln was the only U.S. President to earn a patent. It is very likely that they might have praised the profession for helping make this country great. But as humanitarians, they might have wished we had done more to encourage minorities to carve out careers in engineering and the sciences.

That's why, as we mark Engineers Week in tandem with these Presidents' birthdays, the engineering community is turning a spotlight on diversity. The aim: to give individuals historically underrepresented in engineering and the sciences—blacks, Hispanics, women, and people with disabilities—greater opportunity to develop their intellectual talents to help humankind, not to mention our economy.

Report Shows Only Some Progress

Most engineers love numbers, but the statistics coming out of a 2007 National Science Foundation report titled Women, Minorities, and Persons with Disabilities in Science & Engineering are not uniformly reassuring.

On one hand, there is good news. Hispanics have become better represented in undergraduate engineering programs, and the number of engineering degrees awarded to women has steadily increased every year since 1966. In addition, women went from making up 37% of science and engineering graduate students in 1994 to 42% in 2004.

But there is much work to be done

The National Science Foundation study shows that African American students make up about 6% of engineering undergraduate students. Women's share of bachelor's degrees in computer science dropped from 37% to 25% between 1985 and 2004. Those with physical disabilities also are underrepresented.

The same trends persist in private industry, where minorities, women, and those with disabilities supervise smaller teams of engineers compared with those supervised by whites, men, or those without disabilities.

A Looming Shortage

Why should this worry the private sector? Because engineering talent doesn't come in one ethnicity, color, gender, or physical attribute. Cultivating more technical talent across the board just makes plain, good business sense. The number of retiring workers from science and engineering will mushroom over the next 20 years, aggravating an existing shortfall of technical skills that has already left 1.3 million engineering jobs vacant.

By 2010, the U.S. will need 20% more engineers, yet the growth rate in the number of engineering, math, and science graduates is expected to be about 2%.

There is no better time than Engineers Week and Presidents Day to look to our future—to those we currently refer to as "minorities." By 2050, 85% of workforce entrants are expected to be people of color and women. And, says the National Science Foundation, minorities are expected to make up more than half of the resident college-age population of the U.S. by 2050, up from 34% in 1999. Today's minorities are tomorrow's majorities.

So what to do? For one thing, we ought to think like engineers and apply a healthy dose of persistence and creativity to solve the challenge.

Engineering a Solution

We need greater cooperation between academia, private industry, and government to establish programs that foster enthusiasm and skills for the sciences. Private industry needs to sponsor community mentorships, internships, and workshops for women, minorities, and the physically challenged, and school and camp programs that appeal to them as youths.

The education, hiring, and talent development of these groups ought to be more aggressive as well. We at IBM (IBM) have reaped the fruits of those types of efforts and encourage others to do the same.

Call it a challenge for the social, educational, and economic equivalent of the moon shot, harnessing two of the most powerful forces known to humankind: respect and opportunity. In our efforts this week and year, we would do well to remember that diversity is not an end in and of itself, but a means for achieving equity and cultivating economic opportunity for all.

Washington famously promised to give "bigotry no sanction," and Lincoln encouraged "malice toward none." In short, they believed that thinking people should practice unity and tolerance. Both were reluctantly willing to wage war to accomplish those goals and win freedom for the disadvantaged.

In that same spirit, let's wage an intellectual struggle with the same determination as any physical campaign. As we mark Presidents Day, and as we celebrate the progress we've already made in promoting opportunity within the engineering field, let's fulfill Washington's and Lincoln's ideals, and solve one more Grand Challenge.

John E. Kelly III is an IBM senior vice-president and director of its worldwide research laboratories.

ERIC Identifier: ED433216
Publication Date: 1999-05-00
Author: Clark, Julia V.
Source: **ERIC (Education Resources Information Center)** Clearinghouse for Science Mathematics and Environmental Education Columbus OH.

Minorities in Science and Math. ERIC Digest

While the nation is concerned about the shortage of teachers at the K-12 grade levels, especially in science and mathematics, there is also continuous concern about attracting and retaining more students in these subject areas. Looking to the year 2000 and beyond, we face the potential of a serious shortfall in the number of individuals entering the fields of science and mathematics. This is especially true for underrepresented minority students (Blacks, Hispanics, and American Indians). In the years ahead, these underrepresented minorities will constitute a growing population within the pool of students from which a highly skilled workforce will be drawn.

Minorities are underrepresented at every level from elementary to graduate school. Lack of preparation in science among underrepresented minority groups in the early elementary grades undermines enrollment and success in secondary-level school programs and, ultimately, in college and career choices later in life.

As the nation's economic base shifts increasingly toward technology, participation and achievement in science and mathematics among minority students become increasingly important. Unfortunately, minority students, those who form the most rapidly growing portion of our school-age population, are the ones that are most left out of science and mathematics. By not studying these subjects, both the minority students and the United States as a whole stand to lose. The minority students are depriving themselves of many career choices, including the skilled technical and computer-oriented occupations as well as access to white male-dominated, high salaried occupations. Further, a basic understanding of science and mathematics is essential for all students, not only those pursuing careers in scientific and technical fields. Adequate preparation in science and mathematics enables students to develop intellectually and socially, and participate fully in a technological society as informed citizens (Clark, 1996). The United States can meet future potential shortfalls of scientists and engineers only by reaching out and bringing members of underrepresented minorities into science and engineering. America's standing and competitiveness depend on it (Task Force on Women, Minorities, and the Handicapped in Science and Technology, 1988)

CHANGING DEMOGRAPHICS

Differing fertility rates, immigration patterns, and age distributions, and thus death rates, of population subgroups suggest that the 21st century profile will contrast sharply with that of the 20th century. If the pattern continues, around the year 2030 the total elementary-school-age cohort of the United States could be about equally divided between Whites and all other racial and ethnic groups combined. Over the next 20 years, Blacks, Hispanics, American Indians, and Asian Americans would together outnumber the total White population of elementary school children (Hodgkinson, 1992). The composition of this projected workforce causes great concern in the scientific community and suggests that the United States must make greater efforts in increasing the proportion of minorities choosing careers in science

STATUS OF MINORITIES IN SCIENCE

Too few minorities (Blacks, Hispanics and Native Americans) are represented among the population of scientists in the United States. Despite substantial gain over the past decade, minorities are still underrepresented in science and engineering, both in employment and training (NSF, 1996).

Data from the National Science Foundation (NSF, 1994) indicate that in 1990, racial and ethnic minorities constituted 22% of the civilian labor force but only 14% of the science and engineering labor force. Under-represented minorities (Blacks, Hispanics, and American Indians) represented 19% of the total labor force and 8% of the science and engineering labor force. Asian Americans were well represented in the science and engineering labor force, at 3% of the total labor force and 6% of the science and engineering labor force. Women made up 46% of the labor force in all occupations, but only 22% of the science and engineering labor force.

In the year 2000, it is projected that 85% of new entrants to the workforce in the United States will be females and members of minority groups. Based on this percentage, the goal should be clear. Both groups should be represented in the scientific and technology professions in proportion to their presence in the population as a whole.

Although Blacks demonstrated significant progress during the decade from 1980 to 1990, in both science and math courses taken and in student achievement, they continue to be underrepresented in the science and engineering labor forces. Hispanics also remain underrepresented, with little progress being made during the past decade (NSF, 1994). Limited statistics available on American Indians in the labor force suggest that they too are underrepresented in science and engineering.

Racial inequality persists in Brazil
September 18, 2010

Despite improvements in the past decade, the racial gap in Brazil remains considerably wide, the Brazilian Institute of Geography and Statistics (IBGE) said Friday.

According to IBGE's Summary of Social Indicators (SIS) 2010, black or mixed-race (also referred to as "brown") Brazilians have significantly less schooling and generally earn less than their white counterparts.

Although black and mixed-race citizens made up 51.1 percent of the Brazilian population in 2009, they only accounted for 16 percent of the country's richer class, IBGE said.

Meanwhile, among the poorest Brazilians, black and mixed-race citizens accounted for 9.4 percent and 64.8 percent, respectively.

According to the SIS, Brazil's illiteracy rate among white citizens was 5.9 percent in 2009. But among black and mixed-race citizens, the figures more than doubled to reach 13.3 and 13.4 percent, respectively.

About 15 percent of Brazil's white population aged 15 or older have less than four years of schooling, compared with 25.7 percent among the mixed-race population, and 25.4 percent among the black population. White Brazilians have an average school education of 8.4 years, compared with 6.7 years for the black and mixed-race population.

In addition, the proportion of white Brazilians aged 18 to 24 who attended college reached 62.6 percent in 2009, up from 33.4 percent in 1999. Among their black and mixed-race peers, the figures were only 28.2 percent and 31.8 percent, respectively.

In 2009, only 4.7 percent of the black population and 5.3 percent of the mixed-race population held college diplomas, as opposed to 15 percent among white Brazilians.

The racial gap is also visible in the job market. According to IBGE, black and mixed-race citizens on average earn just 57.4 percent as much as their white counterparts.

Moreover, only 1.7 percent of black and 2.8 percent of mixed-race Brazilians were employers in 2009, compared with 6.1 percent among white citizens.

According to IBGE, the figures showed that the government should pay attention to the improvement of public policies, especially those concerning families with children under 14, which are mostly black or mixed-race.

The IBGE also pointed out that socially vulnerable families formed by single mothers with young children were mostly composed of black and mixed-race citizens.

The government should provide help to these families in order to reduce poverty and improve social cohesion, the IBGE said.